W

Cell Physiologist
Biochemist
and Eccentric

BY
HANS KREBS

in collaboration with
ROSWITHA SCHMID

Translated by
HANS KREBS
AND
ANNE MARTIN

Otto Warburg Cell Physiologist Biochemist and Eccentric

Hans Krebs
with a new
Introduction by Sam Sloan

ISHI PRESS
INTERNATIONAL

Otto Warburg Cell Physiologist Biochemist and Eccentric

Hans Krebs

First Published in German in 1979

Translated into English published in 1981
Reprinted in 2019 by Ishi Press International in New York and Tokyo with a new Introduction by Sam Sloan

ISBN 4-87187-152-5
978-4-87187-152-5
Ishi Press International
1664 Davidson Ave, Suite 1B
Bronx NY 10453
USA
1-917-659-3397
1-917-507-7226
samhsloan@gmail.com

Printed in the United States of America

Otto Warburg Cell Physiologist Biochemist and Eccentric by Hans Krebs Introduction by Sam Sloan

If you have been spending time surfing the Internet you have probably come across an advertisement claiming to have a cure for cancer.

You have probably dismissed it as a hoax and passed it by.

But it is not a hoax. It is however an exaggeration.

They have not discovered a cure for cancer, but they are close.

They have however made progress and we can hope that a cure for cancer will be found soon, in time to save many of us.

But what if people stop dying, where will we live? What if we run out of space to live on?

What if there is not enough medicine to cure all of us? Will there be a committee to decide who will live and who will die?

There are many advertisements for this cure for cancer. Here is one of them:

"Please be warned: The story you're about to

hear was DENIED by our own United States government...
In fact, there is evidence that suggests it has been covered up since the end of the Second World War. Until today...
Even as I speak to you now, certain powerful parties have a vested interest in keeping this information hidden…
…and I fully expect this video to be removed from the Internet in the next 24 hours.
This TRUE story is simply that controversial.
So please listen…
In 1944, at the peak of World War II, did a brilliant German scientist…
…forced to work on the direct orders of Adolf Hitler himself…
…do the “impossible”…
Discover a true CURE for cancer?
This is not hype. I’m talking about what could be the most important discovery in the history of medicine.
And every single detail of this life-saving miracle is preserved… clear as day… in his private journals and papers…
A secret “Nazi treasure” lost for generations... rediscovered…
And finally revealed today.
It’s a fact. Thanks to WWII documents we recently obtained, we now know a German chemist, scientist and medical doctor…
…learned the shocking true CAUSE of all

cancers… a breakthrough that has escaped even today's top scientists…
But this story gets so much bigger… because this German doctor also found the one and only SOLUTION for cancer.
The ultimate "Holy Grail" of medical science...
What's more, the treatment he discovered is so simple and painless, it involves no chemotherapy, radiation or surgery.
And, friend, if you or a loved one has EVER had cancer, you MUST get this solution now."

https://pro.livingwelltoday.live/p/NHS_warburg1yr_1118/LNHSV563/?h=true

Although Dr. Otto Warburg spent most of his life performing cancer research, he did not win the Nobel Prize for it, for reasons that are explained. In 1931, he won the Nobel Prize in Medicine/Physiology "for his discovery of the nature and mode of action of the respiratory enzyme." It was widely rumored among Dr. Warburg's peers that he also won the 1944 Nobel Prize in Medicine/Physiology "for his discovery of the active groups of the hydrogen transferring enzymes." However, in 1937 Adolph Hitler had declared a decree that forbade German citizens from accepting the Nobel Prize. Although the Nobel Foundation had denied Dr. Warburg had been selected for a second Nobel Prize, it probably had done it for political and economic reasons. During World War II, Sweden was a neutral country and benefitted from its neutrality by selling massive amounts of iron to Germany's war machine. During World War II, iron export was Sweden's largest source of income. Regardless of what actually happened, after reviewing his works, Dr. Otto Warburg could had won several Nobel Prizes over his distinguished career.

His cancer research was meticulously detailed in The Metabolism of Tumors, which is book five of five of the "Understand Cancer" series from EnCognitive.com.

More than any person in history, Dr. Warburg laid the foundation for the new field of biochemistry—the biology and chemistry of Life at the molecular and cellular levels. In biochemistry, Dr. Warburg's credentials and achievements are unparalleled.

Otto Warburg: A Life Devoted To Science

Seldom does a week go by without articles, internet posts, and advertisements which maintain that Nobel Laureate Otto Warburg discovered that cancer was caused by low oxygen and acidic pH. Though often misreported, Dr. Otto Warburg did not win the Nobel Prize for finding that cancer cannot live in an oxygen rich or alkaline environment.

What Warburg discovered and what he was awarded a Nobel Prize for was for simply making the discovery that low oxygen was characteristic of cancer cells.

Otto Warburg made some groundbreaking discoveries about the mechanisms of cancer, especially as regards respiration and mitochondrial malfunction. Instead of the widely spread disinformation that Warburg discovered that lack of oxygen was the prime cause of cancer and that he discovered that oxygen was a cure for cancer, the truth may surprise you…

What Warburg actually discovered was that cancer cells were low in oxygen due to a change in cellular respiration from using oxygen to using fermentation of sugar. In his own words:

"Cancer, above all other diseases, has countless secondary causes. But, even for cancer, there is only one prime cause. Summarized in a few words, the prime cause of cancer is the replacement of the respiration of oxygen in normal body cells by a fermentation of sugar."

The widespread misrepresentation of Warburg's discovery is most often used to support the sale of products, books, or treatments which center around beating cancer with oxygen and alkalizing. At one time, Warburg did hypothesize that oxygen might be used to cure cancer.

However, when he tested his hypothesis and tried to cure cancer with oxygen, he failed.

It is easy to see why Warburg's efforts to cure cancer with added oxygen failed. This is because lack of oxygen is not the prime cause of cancer – toxins are the prime cause of cancer. Cancer cells are low in oxygen primarily because they have changed from taking in and utilizing oxygen for respiration to a more primitive form of respiration which utilizes sugar instead of oxygen. It is the cancer process itself which causes most of the lack of oxygen, not the lack of oxygen which causes the cancer process.

Regardless of what Warburg postulated and wrote earlier, in his later years he became convinced that illness resulted from pollution. In that, he came much closer to the true primary cause of cancer and other illness. This is something that had been postulated half a century earlier by the late great French scientist Antoine Bechamp. Bechamp believed that illness is the result of a combination of toxins and an unhealthy cellular terrain.

Prolonged exposure to toxins, especially in combination with cells which have not been properly nourished, oxygenated, hydrated and cleansed is THE primary cause of cancer. This is especially so if one views radiation as a toxin. Over time the stress and inflammation that result from toxins leads to a dysfunction in the cellular mitochondria.

This leads to a cellular defense mechanism in which cells revert to a more primitive form of respiration (sugar fermentation), refuse to die, multiply and form a protective barrier.

Simply stated, cancer cells have low oxygen because they quit taking in oxygen for respiration as part of the cancer process itself. Similarly, it is the cancer process itself which causes the body to become increasingly acidic. The body labors mightily to maintain blood pH within a narrow range (7.35 – 7.45) and most people first get cancer when their pH is at or close to the normal range. As the cancer progresses, the body becomes increasingly acidic.

There have been many "prime causes" of cancer put forward, such as fungus, bacteria, viruses, parasites, stress, low pH, and on and on. But notably, virtually every one of those has been around for thousands and thousands of years. What has not been around and what tracks exactly with the increased incidence of cancer is toxins.

Sam Sloan
Bronx New York
USA
June 17, 2019

Otto Warburg

PREFACE

Otto Warburg was one of the leading scientists of his time. He died on 1 August 1970 at the age of 86. In an age which has produced many giants of science, he stands out as one of the great pioneers of contemporary biology. During a career devoted entirely to research and extending over 60 years, he made an exceptionally large number of highly original and far-reaching contributions to cell biology and biochemistry. Lewis and Randall (45), in the preface to their *Thermodynamics and the free energy of chemical substances*, liken the edifice of science to a cathedral built by the efforts of a few architects and many workers. In this sense Warburg was one of the small band of real architects of his generation.

My association with Otto Warburg began in January 1926 when I joined his laboratory as a scientific assistant. I stayed there until the end of March 1930 and, except for the interruption of the war years, I was in regular contact with him until his death. During our earliest encounters it became clear to me that Warburg was a unique personality and I felt that I ought to keep notes about his way of life and some of his sayings. These 'Warburgiana' proved to be useful in writing this biography; they have helped me to let Warburg speak in his own words.

My personal knowledge of Warburg together with his publications and my interviews with many of Warburg's associates are the main sources of the biography. The text is a translation, with a few minor changes and additions, of my German biography of Warburg, published as Volume 41 of the series *Grosse Naturforscher* by the Wissenschaftliche Verlagsgesellschaft mbH, Stuttgart in 1979. This book was in turn an expansion of the obituary memoir I prepared in English for the Royal Society of London (*Biographical Memoirs of Fellows of The Royal Society*, Vol. 18, November 1972). The expansion was in the main a fuller characterization of Warburg's personality, and photographs were added. I am indebted to the following people for supplying me with the information used in the book.

Family: Gertrud von Wartenberg (sister), Dr. Peter G. Meyer Viol (nephew), Elisabeth Thalgott (niece), Dr. Johannes Fuchs (second cousin), Eric M. Warburg (distant relative).
Close friends: Jakob Heiss (companion and close friend from 1919), Paula Schoeller, Dr. Margaret Boveri, Dr. Rudolph Gruber (childhood friend).
Scientific colleagues: M. von Ardenne, D. I. Arnon, H.-U. Bergmeyer, Th. Bücher, D. Burk, A. Butenandt, J. N. Davidson, F. Dickens, H. Gaffron, K. Gawehn, B. Hess, R. Hill, Joan Keilin, G. Krippahl, S. Lorenz, F. Lynen, T. Terranova, H. H. Weber, F. R. Whatley.
Others: Marie-Luise Rehder, Katharina Saupe, Dr. C. C. Worthington, Dr. B. R. Cookson, B. von der Marwitz, Dr. K. F. Reimers.

The suggestion to expand the English obituary memoir I owe to Dr. Roswitha Schmid who later collaborated with me in the preparation of the German book. I wish to record here my thanks for her invaluable help. I also wish to thank Dr. Hans Rotta of the Wissenschaftliche Verlagsgesellschaft for his advice and help in collecting source material and photographs.

Anne Martin skilfully and effectively collaborated with me on the preparation and editing of this translation and I am deeply indebted to her for innumerable critical and constructive comments.

I am much indebted to Daniel I. Arnon, Robert Hill, and Robert Whatley for their help in the section on photosynthesis, to F. Reimers for permission to quote from the film interview of 1966, and to Jakob Heiss for information on numerous details.

Amongst the material I collected there was much which I felt had to be excluded, partly because the reliability of the information could not be checked, partly because the time was not yet ripe; also, consideration for the feelings of contemporaries still alive demanded restraint. I was also encouraged to be frank by Warburg himself who, in conversation with me, once contrasted the style of obituaries in *The Times* of London with that of the German press. German papers, he said, 'all lie' and their obituaries were often distorting eulogies, occasionally unfairly denigrating. *The Times*, he said, gave a fair and objective picture of a personality, the weaknesses included. His remark was perhaps an echo of Goethe (24): 'In German, if one is polite, one lies.' He might also have been thinking of Margret Boveri's book *Wir lügen alle*. Like Warburg, Margret Boveri lived in Berlin-Dahlem in her later years and she was a long-standing friend of Warburg from 1913 until he died.

This book is addressed not only to biologists and biochemists but also to non-specialists who are interested in biology and biochemistry, and to readers who would like to know something of the personality that lay behind Warburg's great scientific achievements. I hope the book will also be of interest to those to whom his biochemical work is not fully comprehensible, because Warburg's singular personality has always been a source of great fascination, even to those who could not appreciate the fine detail of his research achievements. The biography ought to be of special interest to students in areas on which Warburg had a decisive influence. True, students sometimes comment that because of the enormous amount of current knowledge they have to absorb, they have no time to read about the history of their field. But a knowledge of the historical development of a subject is often essential for a full understanding of its present-day situation. To the specialist biochemist, the more or less complete list of Warburg's publications and the references to other relevant publications should be of value.

So soon after the death of a great scientist, no biographer can do full justice to his subject, nor can he assess the full significance of his achievements. The writer can only attempt to describe facets of the personality and assess the scientific achievements as they have affected contemporary developments. He must leave it to future biographers to complete the picture on the basis of further source material and to assess in more detail and more objectively the consequences of the research achievements in the light of later scientific advances.

The passage of time will, I believe, increase Warburg's status as a scientist because his contributions are of a lasting nature and their full importance will emerge only with time. This is bound to mellow attitudes towards some of Warburg's irritating personal qualities, his fierceness in controversy, his oddities, his occasional lack of feeling, and ruthlessness, which embarrassed and puzzled his contemporaries.

Oxford
1980

HAK

CONTENTS

NOTE ON REFERENCES

The author's references in the text are indicated in parentheses ( ); references to publications by Warburg are in square brackets []; additional notes are indicated by superscript numerals.

LIST OF PLATES

ACKNOWLEDGEMENTS FOR ILLUSTRATIONS

Thanks are due to the following for the reproduction of illustrations: Tita Binz, Mannheim (Frontispiece); Elliott and Fry, London (Pl. 1); Dr. Norman Davidson (Pls. 3 & 4); The Nobel Institute, Stockholm (Pl. 6); Ullstein Bilderdienst, Berlin (Pl. 8); Franz Torbecke, Lindau (Pl. 12); Hans Rotta (Pls. 14 & 22); Landesbildstelle, Berlin (Pls. 17 & 18): Walter Köster, Berlin (Pl. 19); Gesellschaft für Biologische Chemie, Berlin (Pl. 20).

I

EARLY YEARS (1883–1918)

Family and early days

Otto Warburg was born on 8 October 1883 in Freiburg im Breisgau where his father, Emil, was at that time Professor of Physics at the University. Emil Warburg (1846–1931) was one of the leading physicists of his time who made many classical contributions to his subject and formed a large school. An appreciation of his achievements and personality was published by James Franck (22), one of his students (see page 77).

The Warburg family can trace its ancestry back to the year 1559.(9, 32, 57, 61, 63, 68, 69, 70) In that year a formal agreement was made between the city of Warburg, ruled by the Prince Bishop of Paderborn, and Simon of Cassel, representing the Jews of Warburg. The agreement protected the Jews and gave them certain rights. Simon's grandson, Jacob Simon Warburg (died 1636) enjoyed a high reputation as the head of the local Jewish community. The story goes that a grandson of his, Jacob Samuel Warburg, was banker to the Margrave of Hesse-Cassel, who pressed him to abandon his Jewish faith and be baptized. Jacob Samuel refused and migrated to the small town of Altona near Hamburg in 1645. The rulers of this area—the Danish crown—were somewhat exceptional in their liberal attitudes in matters of creed, trade, and commerce (52), and this attracted refugees from other parts of Germany and from Holland.

Otto Warburg was the eleventh direct descendant of Simon of Cassel. In Altona and neighbouring Hamburg the family played an important part in local affairs until Hitler interfered. Descendants spread into many countries but Otto's father, Emil, was born in Altona. Many of Simon's other descendants distinguished themselves in scholarship, business, the arts, and public service, and as benefactors.

The banking house of M. M. Warburg in Hamburg was

established in 1798. The art historian, Aby Warburg (1866–1929), was one of the pioneers in his subject. He assembled a unique art history library, financed by his banker brothers. He renounced his right to a share in the banking business on condition that his brothers would pay the bills for all the books that he deemed necessary for his library.[1] Although the brothers enormously underrated the magnitude of this financial obligation, they fulfilled their commitment faithfully, without demur.

After the advent of the Hitler regime, it was fortunately possible to transfer the library and its staff to London where it formed the nucleus of the Warburg Institute of London University. Sir Siegmund Warburg of London, the distinguished banker, is another member of the family. Sir Oscar Warburg (1876–1937) was a wrangler at Cambridge, a member of the London County Council from 1910–1931, and its Chairman in 1925–26. He was a successful horticulturist, businessman, and public servant. The publisher, Frederic J. Warburg, of Secker and Warburg, also belongs to the family, as did the botanist and Zionist leader, Otto Warburg (1859–1938), who distinguished himself in the field of tropical agriculture.

Otto Warburg's mother was Elisabeth Gärtner. She is spoken of as a woman of great vitality and wit. Otto himself thought that some essential traits of his personality stemmed from his mother and her family. She was of South German stock; her family and ancestry include several people who have distinguished themselves as administrators, lawyers, and soldiers. A brother was a General killed in action in the First World War.

Otto had no brothers but three sisters, Käthe, Lotte, and Gertrud.[2] In 1895, when he was 11, the family moved from Freiburg to Berlin. At the Friedrichs-Werder Gymnasium Otto was a good pupil though he occasionally got up to mischievous pranks, as the following communication from the school to his parents of 14 March 1896 indicates:

> The pupil Warburg (lower fourth) has repeatedly taken part in gross misconduct and he has encouraged fellow pupils to join in. When interviewed he exhibited a regrettable lack of truthfulness. This is doubly regrettable in view of his considerable talents and accomplishments and it is urgently desired that these bad traits are energetically dealt with at home.

This trait was certainly corrected: in adult life he became fanatically concerned with establishing truth, as he saw it, a point which he often laboured in his polemic passages and general comments.

The parental home[3] in Berlin became a stimulating cultural centre which brought young Otto into close contact with leading Berlin academics and also with the world of the arts, for his father was an accomplished pianist who played chamber music in his home. Otto took up the violin but abandoned it when he realized that he could not attain his father's standard of accomplishment. He maintained, however, a life-long love for, and appreciation of, music.

Through his father, Otto got to know Emil Fischer (1852–1919), the leading organic chemist of his day, Walter Nernst (1864–1941), the leading physical chemist and physicist, Jacobus Henricus van't Hoff (1852–1911), the pioneer physical chemist, and Theodor Wilhelm Engelmann (1843–1909), the physiologist from whom he first heard something about photosynthesis. Max Planck and Albert Einstein were other members of the circle.

In 1901, Warburg began his studies of chemistry at the University of Freiburg. As was customary in central Europe, he later moved to another university, Berlin, where he completed his studies with a doctoral thesis under Emil Fischer in 1906. The main subject matter of this work was the preparation of optically active peptides and the study of their enzymic hydrolysis. Fischer himself, though head of a large institute with many teaching commitments, spent most of his working time at the bench, side by side with his research students. He set an example by his style of working, his high standards of reliability and personal integrity on which Warburg modelled himself throughout his life.

Pre-First-World-War period

While still a student, Warburg was already exceptionally ambitious. It was his declared aim to do better than his father—and by all ordinary academic standards his father had done extremely well, reaching the top positions open to physicists in Imperial Germany. When Emil Warburg was appointed to the

Chair of Physics at the University of Freiburg he was 30; nineteen years later he was transferred to the Physics Chair at the University of Berlin, the most coveted appointment in the subject. Otto's ambition, however, did not aim at high office (which though giving status and power, consumed too much time and energy on matters of academic routine); his aim was to make great discoveries. Very early, before he graduated, it became his ambition to make a major contribution to research into cancer and especially to find a cure. Although he did not begin to work specifically on cancer until 1922, much of his earlier work appears in retrospect to have been a preparation for his fundamental attack on cancer.

It was his interest in medical problems which prompted him, after the completion of his doctoral thesis, to study medicine at Heidelberg. In his spare time he worked in the laboratory of the Department of Internal Medicine. Under Ludolf Krehl—a distinguished physician and the author of a standard book, *Pathological Physiology*—this was a great centre of medical research where the relevance of the basic sciences to clinical medicine was appreciated.

Warburg obtained his MD degree in 1911 and remained at Heidelberg until 1914, spending occasional periods at the Naples Zoological Station. During these eight years at Heidelberg, he published some 30 major papers. His research students and collaborators during this period included Otto Meyerhof and Julian Huxley.

His first major independent work, published in 1908, concerned the energetics of growth. It dealt with the changes in the oxygen consumption which take place when a sea urchin egg begins to grow after fertilization. These experiments led to the important discovery that on fertilization the rate of oxygen consumption rises up to sixfold.[6, 8, 10] Warburg was already showing exceptional skill in selecting the right kind of material and in perfecting experimental techniques. He chose the sea urchin egg because the mass of live matter is large in relation to the yolk mass and because the development of the fertilized egg is very rapid so that, as he put it, much happens in a short time. His reasoning was that chemical work is done when living matter arises in the course of growth and therefore the rate of the energy supply must rise. Experiments of previous authors on this problem had been inconclusive. By improving the titrimetric method of oxygen determination developed by Winkler and carefully checking possible sources of errors, his results were convincing and established him as an outstanding investigator.

The link between this work and the later investigations on cancer is obvious: when a normal cell becomes cancerous, it grows excessively, and in 1922 Warburg set out to test whether cancer cells have an increased oxygen consumption (see later).

In his subsequent work, Warburg was primarily concerned with fundamental aspects of cell respiration. It was his aim to unravel the nature of the catalysts in living matter which enable molecular oxygen to oxidize substances in near-neutral solution and at biological temperatures that are completely stable in the presence of oxygen. He also appreciated early that a key problem of biological energy transformations is the question of how the free energy made available by combustion is utilized by muscle and other tissues. Fick (21) had made it clear in 1893 that living cells cannot be heat engines, i.e. engines in which heat is an intermediary stage of energy.

Warburg's approach was guided by the conviction that all processes in living matter obey the laws of physics and chemistry, a view now taken for granted but not generally accepted when he entered the field. In 1897 Buchner had succeeded in extracting yeast fermentation from the cell and thus obtaining fermentation in solution. At the time it was held that chemical processes cannot be studied in the solid or semi-solid state (in which they take place *in vivo*) and that it was necessary, as a first step, to bring them into solution and secondly to isolate the components and prepare them in a pure state. When attempting to solubilize 'respiration', i.e. the oxidation of normal cell constituents by molecular oxygen, Warburg found that the main oxidative processes were always attached to insoluble particles, to what he called 'structures'. In 1913, he described 'grana' from liver cells as the structures to which respiration is bound.[33] We now know that these 'grana' were mitochondria, but it took another 30 years before, in the hands of Claude, Hogeboom, Hotchkiss and Schneider, the morphological structure described in the 1890s as mitochondria were shown to be identical with Warburg's grana and constituted the 'power plant' of the cell.(13) To this day, the key reactions of molecular oxygen are virtually inseparable from insoluble components. They are, as Warburg expressed it, 'structure-bound' or as we now say, 'membrane-bound'. Warburg made many ingenious efforts to characterize the properties of the structure-bound oxidations. He studied in particular how they can be influenced by extraneous substances, hoping that by such effects information could be discovered about the nature of the catalysts of respiration. He established that many substances such as alcohols, nitriles, or urethanes inhibit oxidations according to their surface activity and he concluded that narcotics—substances which are chemically relatively indifferent and inhibit respiration reversibly—interfere because their absorption on to the surface of the oxidizing structures prevents access of substrates to the enzyme. An exception where surface activity was clearly irrelevant was the action of cyanide, a powerful inhibitor of biological oxidations. Since cyanide readily reacts with heavy metals, Warburg rightly concluded that heavy metal catalysts must be involved in cell respiration. He demonstrated the presence of iron in biological material and by 'model' experiments—the acceleration of the oxidation of tartrate, lecithine, linolenic acid, cysteine, and aldehydes by iron salts [35]—he supported the hypothesis that the heavy metal concerned is an iron compound.

Warburg summarized the work carried out between 1906 and 1914 and its general implications in a comprehensive review.[40]

A later generation of scientists occasionally has difficulty in appreciating how Warburg arrived at the concept of 'structure-bound' processes in the cell. Thus in his Nobel Lecture, Luis Leloir (44) wrote in 1971: 'In 1936 we managed to prepare a cell-free system which was active when suitably supplemented and this was a novel result since the process of oxidation was believed to require the integrity of the cells. *I suppose the young generation of biochemists finds it difficult to understand many of the things which we believed at that time.*' [my italics.]

This comment is somewhat misleading because the statement that oxidations in living cells require the integrity of the cell structure was not a matter of belief; it was purely descriptive. It expressed the fact that the capacity to oxidize is lost (or largely lost) when the organized structure of cells is destroyed. It was only in the later 1930s that it became clear why the oxidative capacity (and other functions) were lost, namely because of dilution, or loss of, soluble co-factors, such as the adenine- and pyridine nucleotides. The oxidative capacity was also lost in the first instance in Leloir's experiments when he prepared the liver extract but it was restored by the addition of AMP, phosphate, cytochrome C, and fumarate.

Subsequent developments in cell biochemistry fully justified Warburg's view that many chemical processes in the cell are bound to structures. 'Bound to structure' does not imply that components of the structure cannot be brought into solution; it means that they are, *in vivo,* fixed to a specific space and that, once separated from the structure, the functions of the components become impaired. Structures of this kind are mitochondrial membranes, ribosomes (the organelle of protein synthesis), or the outer cell membrane. Location at a fixed point within a membrane is essential for many complex mechanisms, such as energy transformations. Membranes are also necessary to separate compartments within a cell, and they are not merely 'skins' holding materials together, but are highly organized mosaic structures containing enzymes and specific carriers which regulate the transport of substances into and out of cells or cell compartments. Membranes may also serve as a matrix in which the components of a multi-enzyme complex are embedded in a fixed order.

In 1913, at the age of 30, Warburg was appointed 'Member' of the Kaiser Wilhelm Gesellschaft[4] to take effect on 1 April 1914. This was an appointment as head of a research department, which gave him complete freedom in the choice of his research subject and involved no teaching or administrative responsibilities whatsoever. The main movers in bringing about this appointment were Theodor Boveri, the biologist, and his former teacher, Emil Fischer. Fischer described the attitude of the Kaiser Wilhelm Gesellschaft as follows when in 1911 he (successfully) persuaded Richard Willstätter to accept a similar post:

> You will be completely independent. No one will ever trouble you. No one will ever interfere. You may walk in the woods for a few years or, if you like, may ponder over something beautiful.(73)

On the whole, this policy (which rested on extreme care and competence in selecting the right people) paid magnificent dividends: Warburg, Einstein, Haber, Hahn, Meitner, Correns, Goldschmidt, Polanyi, and many others were 'Members' who made the fullest use of the opportunities.

While Warburg's laboratory was being completed and equipped, he worked in the laboratory of Walther Nernst on oxidation–reduction potentials in living systems, but this work was interrupted by the outbreak of the First World War on 1 August 1914.

War service

Soon after the outbreak of the war, Warburg volunteered for service and applied to what was looked upon as a crack cavalry regiment, the Horse guards ('Garde Ulanen') at Potsdam. He was accepted and soon saw service in the front lines. He rose to the rank of Lieutenant, was wounded in action, and was decorated with the Iron Cross, First Class.

In March 1918 he received a letter from Albert Einstein (seven months before the end of fighting) which is characteristic of both Einstein and Warburg.

In order to appreciate some points of the letter it should be borne in mind that Einstein for many years had been a colleague and a friend of Emil Warburg, Otto's father. It was Emil Warburg who in the 1910s had provided the first experimental proof of Einstein's prediction of the law of photochemical equivalence. Emil Warburg and Einstein also had frequent social contacts because they played chamber music together. So Einstein had met Otto Warburg on many occasions in his parents' home before the letter was written (translated from German).

Dear Colleague,

You will be surprised to receive a letter from me because up to now we have walked round each other, without actually getting to know each other. I even fear that by this letter I might arouse something like displeasure; but I *must* write.

I gather that you are one of the most able and most promising younger physiologists in Germany and that the representation of your special subject here is rather mediocre. I also gather that you are on active service in a very dangerous position so that your life continuously hangs on a thread: now for a moment please slip out of your skin and into that of another clear-eyed being and ask yourself: Is this not madness? Can your place out there not be taken by any average man; Is it not important to prevent the loss of valuable men in that bloody struggle? You know this well and must agree with me. Yesterday I spoke to Professor Kraus who entirely shares my opinion and is also willing to make arrangements for you to be claimed for other work.

I therefore entreat you, as a consequence of what I have said, that you may assist us in our endeavours to safeguard your life. I beg you to send me, after a few hours of serious heart-searching, a few lines so that we may know here that our efforts will not fail on account of *your* attitude.

In the anxious hope that in this matter, as an exception, reason will for once prevail, I am with cordial greetings,

Yours sincerely,
A. Einstein

The letter illustrates the very high esteem in which Otto Warburg was held as a young scientist; he was, after all, not yet 31 when the war had started. It also indicates that he must have had the reputation of being strong-willed and not easily diverted from his views. In style and substance the letter is also typical of Einstein. It shows his humanity and his deep concern for younger people. There are many other instances where he troubled himself to intervene in order to help, as illustrated by the Einstein–Born (17) correspondence.

Incidentally, Einstein's comment on the rather mediocre state of German physiology and biochemistry was perfectly fair, considering that his assessment must have been based on the exceptionally high standing of the exact sciences in Germany at that time, distinguished by such names as Planck, Nernst, Born, von Laue, Wien, Emil Fischer, von Baeyer, Willstätter, Wieland, Ostwald, Wallach, Hans Fischer, Haber, Hilbert, and Klein.

Warburg accepted Einstein's proposal because, as he said, the war was in any case lost, and he returned to Berlin soon afterwards. Perhaps Einstein's intervention saved his life.

On his service in the First World War Warburg himself commented in a filmed interview recorded in 1966 by the Institut für den Wissenschaftlichen Film of Göttingen:

The only interruption in my research during the past 53 years was four years in World War I. I do not regret this interruption. In one of the finest uniforms of the old Prussian Army I rode many patrols in advance of the front line during the early advances in Russia. Later when the war of movement had ended I was Orderly Officer to several of our great army commanders. In the course of this I got to know the realities of life which had escaped me in the laboratory. I learned to handle people; I learned to obey and to command. I was taught that one must be more than one appears to be.

2

MAIN SCIENTIFIC ACHIEVEMENTS (1919–1970)

General aspects

After his return to civil life Warburg's laboratory occupied a section of the top floor of the Kaiser Wilhelm Institut für Biologie in Berlin-Dahlem until 1931. It was not a large laboratory, providing initially benches for about six people, and after separate animal quarters were provided in 1927 there were about ten places. Equipment was excellent. Throughout his life it was Warburg's policy to keep the number of research workers low and to use a relatively large portion of his financial resources for technical equipment. His long-term collaborators throughout were technicians who in the 1920s occupied four of the few places—E. Negelein, F. Kubowitz, E. Haas, and W. Christian, to be joined later by W. Lüttgens. He recruited these technicians from skilled mechanics trained in the workshops of engineering firms. Their asset was that they knew how to deal with physical instruments. Warburg himself trained them in chemistry. Many of the technicians, including those mentioned above, became well-known through their publications. Technicians recruited after the Second World War (when Warburg had to start again from scratch) included G. Krippahl, K. Gawehn, H.-S. Gewitz, W. Völker, A. Lehmann, K. Jetschmann, S. Lorenz, A.-W. Geissler, W. Schröder, H. Klotzsch, H. Geleick, and H.-W. Gattung. Warburg found technicians ideal collaborators because they were content with their status and, unlike academics, were not liable to focus their interests on promotion and careers: they never troubled him with requests for testimonials nor for support to obtain better posts. He treated them well, but firmly. Their salaries were appropriate to their long working hours. They were devoted to him and stayed many years.

Post-doctoral collaborators included H. Gaffron, D. Burk,

W. Kempner, W. Cremer, H. Theorell, E. G. Ball, C. S. French, J. N. Davidson, Th. Bücker, and myself. The total number of research workers in his laboratory at any one time was usually no more than 10–12 of whom perhaps one or two were academics. (This figure refers to the time after his move to a new Institute in 1931.)

In 1929 Warburg gave a review lecture—a Herter Lecture—at Baltimore entitled 'Enzyme action and biological oxidation'. He recorded [435] that after this lecture the Rockefeller Foundation offered to support his research. He suggested the building of a small institute for cell physiology, as well as the foundation of a larger institute for physics under the direction of Max von Laue. He had in mind collaboration with physicists, especially in the field of radiation, the application of which to biochemical problems had been a major source of his recent successes. Within six months the Rockefeller Foundation provided 635 000 marks for the purchase of the land required for the two institutes, 600 000 marks for the building and equipment of the institute for cell physiology and about 1.5 million marks for an institute of physics, all under the wing of the Kaiser Wilhelm Gesellschaft.

The new institute—the Kaiser Wilhelm Institut für Zellphysiologie—was ready for occupation late in 1931. Warburg chose as a model for its outer style the fine eighteenth-century rococo manor house at Grosskreutz near Potsdam, owned by the von der Marwitz family.[5] This unusually attractive building provided all the space he needed, including facilities for the large-scale chemical operations necessary for the initial steps of enzyme isolation.

In 1943 air attacks made life in Berlin dangerous and Warburg, his staff and equipment moved to the Liebenberg estate, about thirty miles north of Berlin, which had been placed at his disposal by the owner, Prince Eulenberg, a Prussian nobleman. Here work continued undisturbed until 1945 when the advancing Russians occupied the area and removed all the equipment from the laboratory. Who was responsible for this action has never been clarified, but afterwards the Russian Commander-in-Chief, Marshal Zhukov, invited Warburg to see him and told him, in the name of the Russian Government, that the dismantling of his laboratory had been an error. The Marshal issued orders for the apparatus and books to be returned but, alas, they could not be traced.

The Dahlem Institute survived the war undamaged. It was located in the American Sector of Berlin and for over four years had been used by the American forces as the headquarters of their Berlin High Command. They evacuated the building in 1949 and with help from many quarters it was, in Warburg's own words, 'reconverted into the Kaiser Wilhelm Institut für Zellphysiologie'.[435] It was ceremonially re-opened on 8 May 1950, by General Maxwell D. Taylor, the Commandant of the American Sector of Berlin and later Chief Military Adviser to President John F. Kennedy. In 1953 the Institute was renamed Max Planck Institut für Zellphysiologie.[6]

Thus for five years Warburg was prevented from carrying out experimental work but during this period he wrote two books [506, 507], in which he summarized his earlier work and discussed its significance, especially in relation to the work of other investigators. In 1948–49 he spent several months in the United States, at Urbana, Bethesda, and Woods Hole.

Within a few years of Warburg's return from the First World War, three main lines of investigations emerged—photosynthesis, cancer, and the chemical nature of the enzymes responsible for biological energy transformations, i.e. of oxidations and reductions. These three subjects (in many ways interlinked) with their numerous ramifications occupied him throughout his life from 1919 to 1970. In all three he greatly advanced the methodology and made fundamental discoveries.

Advances in methodology

Manometry

In his early studies Warburg had used titrimetric methods for measuring the oxygen uptake of sea-urchin eggs and red blood cells but these procedures were cumbersome and not very sensitive. Having seen the Haldane–Barcroft 'blood gas manometer' during a brief visit to Barcroft's laboratory at Cambridge,[7] he began to use this instrument in 1910 when he referred to it as the 'beautiful blood gas apparatus of Haldane–Barcroft'.[11] This differed from earlier manometers in that a

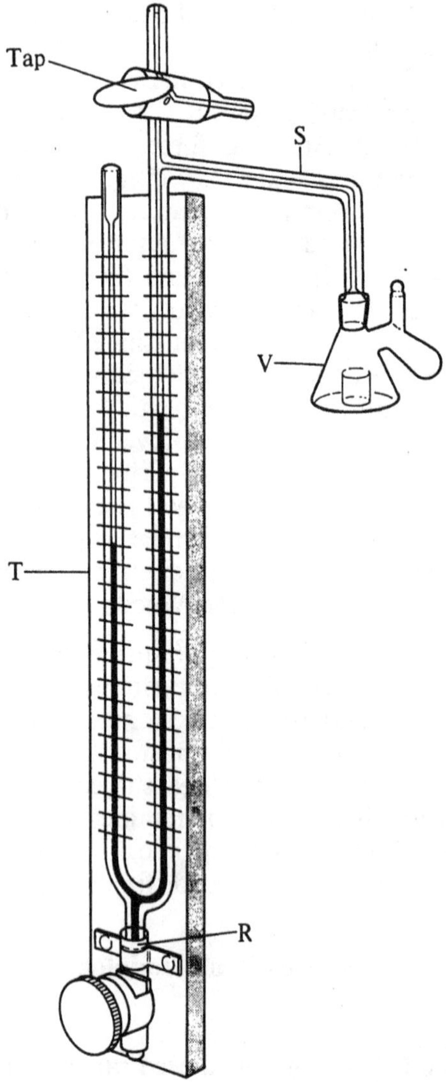

Fig. 1. The Warburg manometer consists of a U-tube T of narrow bore, graduated in millimetres. At the bottom the tube has a rubber reservoir R and a screw-clamp arrangement by which the level of the liquid in the tube can be adjusted. The left arm of the tube is open to the air; the right arm carries a side arm S to which glass vessels V (of various shapes) can be attached by means of a ground joint. On the top of the right-hand arm is a tap, by which the vessel can be closed or opened to the air. The manometer is mounted on a board which can be attached to a shaking apparatus.

special device kept the volume of the gas space constant so that pressure was the only variable when a gas was formed or removed at constant temperature. The originators had used the apparatus for measuring the *quantities* of oxygen bound to haemoglobin and for the carbon dioxide content of blood. Warburg adapted it to measure *rates* of gas exchange. For this purpose it was necessary to maintain gas equilibrium between the liquid and the gas phases in the manometer vessel and this was achieved by constant shaking of the vessels in a water bath. [57, 64, 77]

In many situations manometry proved far more accurate and far more convenient than the classical methods of gas analysis. It was applicable not only to processes in which gases participate directly, as in cell respiration, photosynthesis, and

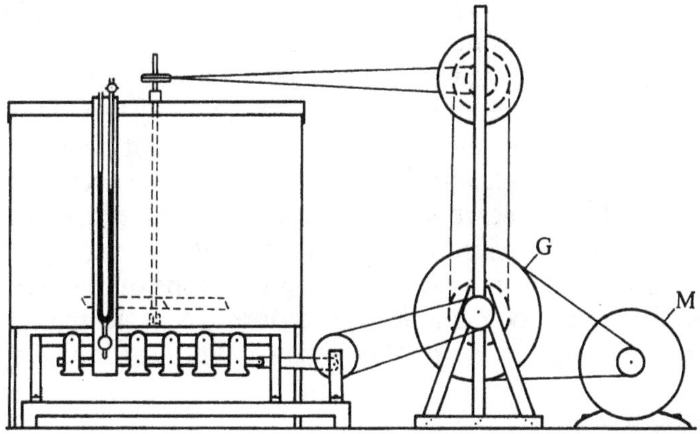

The manometer measures pressure changes in the vessel at constant temperature and volume. Readings are taken of the level differences between the right and left arms, the level in the right arm being adjusted to a zero point by the screw at the bottom of the manometer. By this adjustment to zero, the volume of the vessel is kept constant. When temperature and volume are constant, the change in pressure is directly proportional to the amount of gas evolved or absorbed.

The calculation is based on the gas laws. This is possible if equilibrium is maintained between gas phase and liquid phase within the vessel. The shaking of the manometer and the stirring of the bath maintain equilibrium at constant temperature. The shaking speed can be varied by a system of belts and pulleys G driven by an electric motor M.

A detailed description of the theory and practice is given in *Manometric methods* by Malcolm Dixon (Cambridge University Press, 1943).

alcoholic fermentation, but also to reactions (such as lactic acid production by animal tissues) which can be coupled with a gas-producing reaction by allowing them to take place in the presence of bicarbonate, the acid produced leading to the liberation of an equivalent amount of CO_2.

A special advantage of the manometric technique developed by Warburg, as opposed to other manometric or gasometric methods (for instance those of Van Slyke and of Haldane), is the fact that it measures increments directly whilst the others measure differences. Hence the presence of any say, O_2 or CO_2 does not affect the accuracy of the measurements.

The theory and practice of manometry were perfected in the 1920s, but further important elaborations devised by Warburg were added in the 1950s.[508] Manometry was the key factor in the discovery of the lactic acid fermentation of cancer tissue, in much of the work on cell respiration, fermentation, and photosynthesis, as well as in the identification of the iron porphyrin structure of the oxygen-transferring enzyme of cell respiration. The manometric method also proved exceedingly valuable as an analytical tool for the quantitative determination of small amounts of, for example, bicarbonate, urea, succinic acid, amino acids (by decarboxylation reactions), and purine bases.

A 600-page monograph *Manometrische Methoden* (40) edited by A. Kleinzeller in 1966 describing the scope of manometry, including over 100 analytical procedures, is indicative of the value and range of the methods which Warburg initiated.

Spectrophotometry

In the 1940s, some aspects of manometry began to be superseded by spectrophotometric techniques, but this again was the result of Warburg's fundamental discoveries and methodological ingenuity. It was Warburg who developed from 1928 onwards the use of the photoelectric cell and the provision of suitable sources of monochromatic light, in the first instance for his measurement of the action spectrum of the 'O_2-transferring enzyme of respiration'.[129, 140] Much later, in the 1940s, these principles were incorporated into commercial spectrophotometers, the first to be marketed being that of the Beckman company. It was Warburg who discovered the absorption of reduced pyridine nucleotides at 340 nm [253], and who used this property as the basis of a large number of 'optical tests' for measuring reaction rates and quantities of metabolites. He furthermore established the principle that the extinction of reduced pyridine nucleotide at 340 nm can be used for the measurement of reactions and substances which are not directly involved in oxido-reductions. Coupling reactions can establish a link with dehydrogenase systems. Thus aldolase activity can be determined by coupling with glyceraldehyde-phosphate dehydrogenase so that a product of the aldolase reaction undergoes dehydrogenation. The optical tests were also essential tools in the purification and eventual crystallization of numerous enzymes. Requiring minute quantities of material and usually taking merely minutes to perform, they indicate in which fraction the active enzyme protein is contained.[8]

Tissue slice technique

The use of tissue slices was developed by Warburg in the first instance for the measurement of the respiration of cancer cells because he wanted to test whether oxygen uptake increases when cells begin to grow.[57] The object of slices was to have a sufficiently thin piece of tissue where diffusion from and to the suspending medium adequately supplies nutrients and removes waste products, a function normally effected by the blood circulation. Slices less than 0.5 mm thick are shaken in a saline medium resembling blood serum in respect to inorganic constituents, or in serum itself. Warburg developed the mathematical theory necessary for the calculation of the maximum thickness permissible to saturate the tissue with oxygen; this value depends on the rate of oxygen consumption, on the rate of diffusion within the tissue and on the concentration of O_2 in the medium. Slices usually contain 100–150 layers of cells, and as only one cell on each side is cut and directly damaged by the razor, the structure of the majority of cells is expected to remain intact. Intactness of cell structure was of importance because Warburg's earlier work (already mentioned) had shown that the energy-giving reactions are lost when the cell structure is disrupted. Thus it was necessary to devise a method for measuring the energy-giving processes without mechanical destruction of the tissue and without addition of antiseptics, points which the earlier investigators did not adequately appreciate. True, it has since become possible to restore the energy-giving reactions of homogenized tissues, but only by the addition of co-factors, and the rates observed in such preparations reflect the *in vivo* rates much less reliably than those found in slices.

It is a decisive advantage of the slice technique that samples of the same organ can be incubated in parallel under a variety of precisely controlled conditions. In the case of most tissues, 50 to 100 mg fresh weight are sufficient for accurate measurements of oxygen consumption and lactate production. An alternative and already well-established method was the perfusion of the isolated intact organs, of which Warburg had personal experience.[13] This method was applicable to organs like liver, kidney, or heart from larger animals, which have a clear-cut vascular system through which they can be perfused. Cancer tissue, as a rule, has an irregular blood supply and therefore does not lend itself to perfusion experiments. Moreover, the perfusion technique does not allow parallel experiments to be carried out on the same tissue, and it could not (yet) be applied to small animals. It was therefore cumbersome and expensive.

The slice technique was originally designed to study the energy-giving processes. Later it became clear that slices also perform many biosynthetic reactions and they proved an invaluable material in the study of intermediary metabolism, both degradative and biosynthetic, and also in the study of the mechanisms responsible for the transport of substances into and out of tissues. The tissue-slice technique thus proved of great importance generally to the study of metabolic processes.

Cancer, tissue metabolism, and the Pasteur effect

The cancer problem, as already mentioned, began to occupy Warburg's mind in his student days when he became aware of the ravages of cancer and of the great limitations of successful treatment. His approach was to be a fundamental one—an attempt to find out what biochemical changes take place in a tissue when a normal cell (the growth of which is controlled) becomes a cancer cell (the growth of which is unrestricted). Does the metabolism of cancer cells differ qualitatively from that of normal tissues? The question had been asked before but earlier attempts to find an answer, in Warburg's view, suffered from both conceptional and experimental shortcomings. If the abnormal growth rate had to be explained, he argued, one would have to consider in the first instance the reactions which provide energy for growth because without energy, growth cannot take place. He had in mind his earlier discovery that the respiration of the sea-urchin egg can increase up to sixfold when the egg is fertilized and begins to grow, and he therefore set out in 1923 to measure the rates of respiration of a transplantable cancer—the Flexner–Jobling rat carcinoma—using the tissue-slice technique developed for this purpose.[58, 66, 504] The results were clear-cut. Firstly, the rate of oxygen consumption of the cancer cells did not differ from that of a variety of normal cells. Secondly, cancer cells readily form lactate from glucose to an extent which far exceeds the rates of liver, kidney, pancreas, and submaxillary gland. The rate was high enough to make a significant contribution to the energy supply of the tissue. These findings were soon confirmed for other types of neoplastic cells, including human cancers. A special feature of the high rate of 'glycolysis' of cancer cells was its occurrence in the *presence* of oxygen. It had already been known before Warburg that many tissues, e.g. muscle, can form lactate from carbohydrate in the *absence* of oxygen.

Warburg asked himself at once whether the aerobic glycolysis is specific for neoplastic cells and he therefore examined many tissues systematically for their capacity to glycolyse aerobically and anaerobically. This led to the discovery that all animal tissues that are metabolically active glycolyse anaerobically and that the great majority of these do not glycolyse aerobically. The main exception was the retina of warm-blooded animals, the aerobic glycolysis of which was even faster than that of cancer tissue. Furthermore he found that some tissues, for example the haemopoietic bone marrow cells, gradually develop an aerobic glycolysis when they are exposed to unphysiological conditions *in vitro*. This he took to indicate that the aerobic glycolysis of the retina, a very delicate tissue, might be an *in vitro* artefact.

Warburg related these discoveries to analogous observations which Pasteur had made 60 years earlier in micro-organisms. Pasteur established that the rates of fermentation are generally high anaerobically, but low aerobically. It was in fact Pasteur who coined the terms aerobic and anaerobic. So Warburg reached the conclusion that cancer cells differ from non-cancer cells, including growing embryonic cells, by their failure to suppress glycolysis in the presence of oxygen.

By this time, Meyerhof had already established that in muscle there is a regular relationship between the rate of respiration and the suppression of glycolysis: usually the consumption of one molecule of oxygen prevents the formation of the two molecules of lactate. This meant that the effect of oxygen cannot be explained (as previous investigators had thought) by a complete oxidation of lactate. Meyerhof suggested that many findings agreed with the assumption that the enzymes forming lactate also function aerobically but that the lactate formed is resynthesized to carbohydrate at the expense of the energy provided by respiration. He depicted the carbohydrate metabolism of muscle by this diagram:

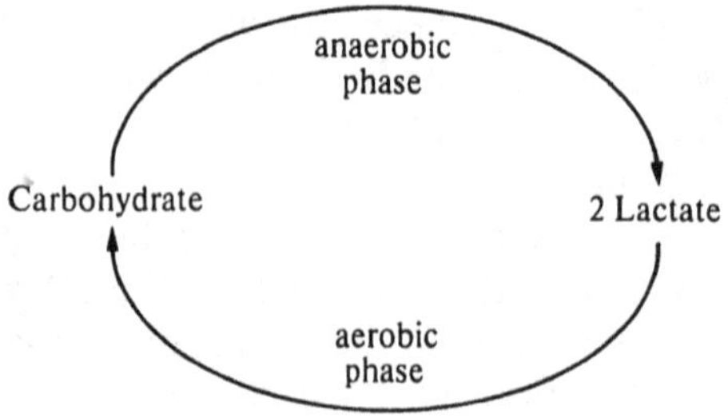

He calculated that sufficient energy would become available from the consumption of one molecule of oxygen to bring about a resynthesis of carbohydrate from two molecules of lactate.

Having discovered the high aerobic glycolysis of cancer cells, Warburg attempted to establish the reason for the failure of oxygen to suppress glycolysis. His measurements of the anaerobic and aerobic lactic acid production by various tissues, and the comparison with the measurements of the oxygen consumption, showed that in many tissues approximately two molecules of lactate are prevented from appearing when one molecule of oxygen is taken up by respiration. This is the same quantitative relation which Meyerhof had found in muscle tissue. To express the effectiveness of respiration in preventing aerobic glycolysis, Warburg introduced the quotient

$$\frac{\text{anaerobic glycolysis} \quad \text{minus} \quad \text{aerobic glycolysis}}{\text{respiration}}$$

and called it the 'Meyerhof quotient'. When glycolysis is expressed in units of lactate formed and respiration in terms of units O_2 consumed, the value of the Meyerhof quotient is usually two. In the case of cancer tissue he found respiration either less efficient or altogether smaller, and thus reached the conclusion that in cancer cells respiration is defective in controlling aerobic glycolysis.

At this stage it was entirely obscure by which kind of mechanism oxygen might act in preventing the products of fermentation from appearing and in bringing about a resynthesis of carbohydrate. The time was not ready because the pathways of glycolysis and gluconeogenesis had not yet been clarified. A contribution to the subject of the relations between respiration and fermentation was Warburg's discovery in 1926 that the link between respiration and fermentation can be severed by a specific inhibitor, ethylcarbylamine.[87] He found that this substance, at low concentration, does not inhibit the respiration of many cells, yet at the same concentrations cells exposed to ethylcarbylamine glycolysis in oxygen at almost the same rate as in the absence of oxygen. In other words, ethylcarbylamine 'stimulated' aerobic glycolysis. On the basis of this finding Warburg visualized a link between respiration and fermentation in terms of a specific chemical reaction and he introduced the term 'Pasteur reaction' for this chemical link. He looked upon ethylcarbylamine as a specific inhibitor of the Pasteur reaction. Since ethylcarbylamine readily chelates with heavy metal ions and inhibits other reactions catalysed by heavy metals, Warburg suggested that the catalyst promoting the Pasteur reaction contained a heavy metal. The nature of his hypothetical reaction and its mechanism of action remained obscure for many years in spite of numerous efforts to elucidate it.

Subsequent work initiated by Engelhardt and elaborated by

Lynen, Bücher, Lowry, Racker, and Sols made it clear that this concept of the mechanism of the Pasteur effect is not correct. (41) In fact oxygen does not cause a re-synthesis of the products of fermentation. Respiration, through the synthesis of ATP, modifies the catalytic properties of the rate-limiting enzyme of glycolysis, phosphofructokinase. The activity of this enzyme is variable and is regulated by the concentrations of ATP, ADP, and inorganic phosphate, as well as by various other metabolites. Phosphofructokinase is an 'allosteric' enzyme, and the Pasteur effect can now be accounted for by the allosteric properties of this enzyme: ATP inactivates phosphofructokinase while ADP and inorganic phosphate activate it. Since respiration, through oxidative phosphorylation, causes a conversion of ADP and inorganic phosphate to ATP it decreases the activity of phosphofructokinase; further, since ethylcarbylamine, as was eventually established, uncouples oxidative phosphorylation it can abolish the effect of respiration on glycolysis. Many other 'uncouplers' were found to have the same effect. Thus the 'Pasteur reaction' is an aspect of oxidative phosphorylation. Since heavy metals are components of the respiratory chain, Warburg was right in postulating that heavy metals play a part in the Pasteur reaction.

This answer to the question of the nature of the Pasteur effect was established some 40 years after Warburg had first stated the phenomenon in clear terms; it is now evident that this answer could have been established only after much collateral progress in enzymology had been made.

Warburg's starting point in studying cancer, it will be recalled, was the question whether respiration rises (as it does in the sea-urchin egg) when growth begins. What emerged was that there was no rise in respiration but that there was a different source of energy, a significant amount of energy in neoplastic tissue being derived from glycolysis, even under aerobic conditions. In ascites tumour cells this can be 50 per cent but the *total* energy supply, in terms of ATP production, is no greater in cancer cells than in their non-growing parent cells.

Why, then, is the rate of energy supply unchanged when a normal cell becomes cancerous and begins to grow, whilst there is a great difference between the rates in the non-growing and the growing sea urchin egg? The reason may be this: the unfertilized non-growing egg has an exceedingly low metabolic rate because cell activities are minimal. On the other hand, the parent cells of cancer cells are usually very active. When they become cancerous the energy supply is diverted from the performance of their normal function to growth. This is an aspect of 'de-differentiation', or 're-orientation of gene expression'. Because of this switch-over from one activity to another no increase occurs in the over-all rates of energy supply. In the unfertilized egg there are no significant supplies of ATP which can be redirected; hence the increase on fertilization.

Warburg maintained a keen interest in cancer throughout his life. Following up his earlier work he studied in the 1950s and 1960s how the aerobic glycolysis of cancer cells—implying cancer itself—arises. He extended the work of Goldblatt and Cameron (27) who had found the fibroblasts in tissue culture develop into fibrosarcoma cells when repeatedly exposed for short periods to low oxygen pressure. Warburg showed that embryonic mouse cells aquire the metabolic characteristics of cancer cells in tissue culture at low oxygen pressure within 48 hours, i.e. in the course of two cell divisions. Most important of all, if normal oxygen pressure is restored, the cancer metabolism remains and supports growth. Thus the transformation of embryonic cell metabolism into cancer metabolism can be irreversible, just as the formation of cancer cells from normal body cells is irreversible.(16, 19, 20, 23) On injection into the animal the transformed embryonic cells are well-known to develop into fatal cancers. Such observations led Warburg to the conviction that the loss of normal respiration is a key factor in carcinogenesis.

He was also very much concerned about the indifference of the authorities and the general public towards the prevention of cancer. He pressed on many occasions for measures which would make proper use of the scientific information. He wrote [477, 494]

> Many experts will agree that one could prevent about 80% of all cancers if one could keep out the known carcinogens. This might involve very little expense and, in particular, would require little further research.[9]

In 1954 he expressed the hope that cigarette smoking, food

additives, and air pollution by motor car exhausts would be greatly restricted. He was indeed exasperated that so little was being done in this direction and that people continued to argue that too little was known about cancer to take effective preventive action. While no one will quarrel with his feelings, his views on the significance of the metabolic characteristics of cancer cells were not shared by the majority of experts—though none of the facts on which they were based have been refuted.

In 1967 he summarized his views in a privately printed paper 'The prime cause and prevention of cancer' [477, 494] which included the following passages.

There are primary and secondary causes of diseases. For example, the primary cause of plague is the plague bacillus, but secondary causes of plague are filth, rats, and the fleas that transfer the plague bacillus from rats to man. By a prime cause of disease I mean one that is found in every case of the disease.

Cancer, above all other diseases, has countless secondary causes. Almost anything can cause cancer. But, even for cancer, there is only one prime cause. Summarised in a few words, the prime cause of cancer is the replacement of the respiration of oxygen in normal body cells by a fermentation of sugar. All normal body cells meet their energy needs by respiration of oxygen, whereas cancer cells meet their energy needs in great part by fermentation. All normal body cells are thus obligate aerobes, whereas all cancer cells are partial anaerobes. From the standpoint of the physics and chemistry of life this difference between normal and cancer cells is so great that one can scarcely picture a greater difference. Oxygen gas, the donor of energy in plants and animals is dethroned in the cancer cells and replaced by an energy yielding reaction of the lowest living forms, namely, a fermentation of glucose.

The key to the cancer problem is accordingly the energetics of life, which has been the field of work in the Dahlem institute since its initiation by the Rockefeller Foundation. In Dahlem the oxygen-transferring and hydrogen-transferring enzymes were discovered and chemically isolated. In Dahlem the fermentation of cancer cells was discovered decades ago; but only in recent years has it been demonstrated that cancer cells can actually grow in the body almost with only the energy of fermentation. Only today can one submit, with respect to cancer, all the experiments demanded by Pasteur and Koch as proof of the primary cause of a disease. If it is true that the replacement of oxygen-respiration by fermentation is the primary cause of cancer, then all cancer cells without exception must ferment, and no normal growing cell ought to exist that ferments in the body.

Most experts agree that nearly 80% of cancers could be prevented, if all contact with the known exogenous carcinogens could be avoided. But how can the remaining 20%, the endogenous or so-called spontaneous cancers, be prevented?

Because no cancer cell exists, the respiration of which is intact, it cannot be disputed that cancer could be prevented if the respiration of body cells could be kept intact.

Today we know two methods by which we can influence cell respiration. The first is to decrease the oxygen pressure in growing cells. If it is decreased to a level where the oxygen-transferring enzymes are no longer saturated with oxygen, respiration can decrease irreversibly and normal cells can be transformed into facultative anaerobes.

The second method of influencing cell respiration *in vivo* is to add the active groups of the respiratory enzymes to the food of man. Lack of these groups impairs cell respiration and abundance of these groups repairs impaired cell respiration—a statement that is proved by the fact that these groups are necessary vitamins for man.

To prevent cancer it is therefore proposed first to keep the speed of the bloodstream so high that the venous blood still contains sufficient oxygen; second, to keep high the concentration of haemoglobin in the blood: third, to add always to the food, even of healthy people, the active groups of the respiratory enzymes: and to increase the dose of these groups, if a pre-cancerous state has already developed. If at the same time exogenous carcinogens are rigorously excluded, then much more endogenous cancer might be prevented today.

These proposals are in no way Utopian. On the contrary, they may be realised by everybody, everywhere, at any hour. Unlike the prevention of many other diseases, the prevention of cancer requires no government help, and not much money.

These passages on the 'primary' cause of cancer, written at the age of 83, still show Warburg's clear, logical, and forceful style but the balance of judgement, in the view of most experts, is at fault. His sweeping generalizations spring from gross simplification. The partial replacement of respiration by glycolysis is only one of many characteristics which distinguish cancer cells from normal cells. Warburg neglected the fundamental biochemical aspect of the cancer problem, that of the mechanisms which are responsible for the controlled growth of normal cells and which are lost or disturbed in the cancer cell. No doubt, the differences in energy metabolism discovered by Warburg are important, but however important, they are at a level of the biochemical organization of the cell, not deep enough to touch the heart of the cancer problem, the uncontrolled growth. Warburg's 'primary cause of cancer'—the replacement of respiration by fermentation—may be a symptom of the primary cause, but is not the primary cause itself. The primary cause is to be expected at the level of the control of gene expression, the minutiae of which are unknown though some of the principles involved are understood.

The catalysts of the respiratory chain

The earlier work on the nature of the catalysts of biological oxidations had convinced Warburg that heavy metals played a key role. He had demonstrated the presence of iron in respiring cells; he had demonstrated the inhibition of biological oxidations by traces of cyanide which combine with iron compounds; and he had established 'models' of respiring systems (e.g. blood charcoal) which simulate biological oxidation in that they catalyse the oxidation by molecular oxygen of physiologically occurring stable substances and also established that this oxidation is sensitive to cyanide. In 1926 he added a new piece of evidence in support of this conclusion which subsequently turned out to be decisive. Since some iron compounds, for example haemoglobin, readily react with carbon monoxide, he tested the effect of carbon monoxide on the respiration of yeast cells and discovered that this substance indeed inhibits respiration, and that the degree of inhibition depends on the oxygen pressure.[115, 128] Hence at a given carbon monoxide pressure, the inhibition is greater the lower the oxygen pressure, a behaviour exactly analogous to the competition of haemoglobin for oxygen and carbon monoxide. While these experiments were in progress in the winter of 1927–28, A. V. Hill visited the laboratory and drew Warburg's attention to the light-sensitivity of carbon monoxide haemoglobin discovered by Haldane and Smith in 1896.(28) A test to find out whether the carbon monoxide compound of the respiratory catalyst was also light-sensitive was carried out immediately as the necessary tools were at hand, and showed that the inhibition of respiration was greatly diminished when the yeast suspension was illuminated. This phenomenon provided Warburg with the possibility of establishing the spectrum of the catalyst and this he exploited with exquisite skill and ingenuity. He illuminated the yeast suspensions with monochromatic light of known intensities and measured quantitatively the effect of light on the inhibition of respiration by carbon monoxide. From these measurements he obtained the absorption spectrum of the catalyst which competes for oxygen and carbon monoxide.[140, 148] The principle involved was simple: since only light absorbed by the catalyst can be effective in removing the inhibition, the absorption spectrum of the catalyst is related to the effectiveness of the light. But to devise and to carry out the experiments and to develop the mathematical analysis of the measurements required very exceptional experimental and theoretical skill. First he had to find sources of monochromatic light of sufficient intensity, then he needed methods for measuring the gas exchanges and light intensities, and finally he had to elaborate the theory for the quantitative interpretation of the measurements. The spectrum found agreed quantitatively with those of iron porphyrins such as haemoglobins and the cytochromes, but it was not identical to either. Further, neither haemoglobins nor cytochromes are oxidizable by molecular oxygen and in this respect differ from Warburg's compound, the 'oxygen-transferring enzyme'. It was this work for which Warburg was awarded the Nobel Prize for Medicine and Physiology in 1931.

The discovery of the CO-sensitive iron porphyrin (or 'haem'), and the effects of light upon it, had to be reconciled with Keilin's discovery of the three cytochromes *a*, *b*, and *c* in 1925, three haem compounds which in living cells readily undergo oxidation and reduction, the oxidation being inhibited by CO, and the reduction by narcotics. These cytochromes showed no change in their absorption spectrum on addition of cyanide or CO, nor was their spectrum identical with Warburg's haem compounds. This was the origin of a prolonged controversy between Warburg and Keilin. [154, 506] Warburg argued that the cytochromes are not directly involved in the interaction between O_2 and the substrates of respiration. The details of this controversy cannot be discussed here and the reader must be referred to the *Biographical memoir* of Keilin (50), Keilin's own record in his book *The history of cell respiration and cytochrome* (36), and to an obituary notice by E. C. Slater.(65) The apparent contradictions were resolved by Keilin's discovery in 1939 (37) that cytochrome *a* consisted of (at least) two components, one of which, named a_3 or cytochrome oxidase, combines with CO, HCN and had the same spectrum as Warburg's 'respiratory enzyme' of 1926. In 1927, Keilin had already shown that the 'indophenol oxidase' (an enzyme of wide occurrence in living material which catalyses the formation of indophenol blue in the presence of α-naphthol and air) is inhibited by CO and by cyanide and that its inhibition by CO is reversed by light. He concluded that indophenol oxidase and Warburg's enzyme were identical and that this enzyme was responsible for the oxidation of the cytochromes. In 1929 Keilin (38) wrote,

> the term 'indophenol oxidase' must be retained in preference to the term 'respiratory enzyme', which is too wide and which implies that it is the only enzyme taking part in the respiratory mechanism of the cell. That is, however, not the case, indophenol oxidase in the living actively respiring cell represents only one link in the chain forming the complicated respiratory mechanism, in which several other respiratory enzymes and systems are involved and intimately connected.

Warburg resented the suggestion that the iron porphyrin discovered by him should be named after an artificial substrate which it happens to be able to oxidize, when in fact it is a catalyst essential in the vast majority of aerobic organisms for the combustion of organic substances by molecular oxygen.[10]

Warburg's conclusion, arrived at in 1914 [35], that iron is a catalyst of cell respiration subsequently proved to be valid to a much wider degree than Warburg could have visualized at the time. In the 1960s it became clear that apart from a series of iron porphyrins, several (at least seven) 'non-haem' iron compounds are involved in electron transport. In these compounds the iron is attached to the sulphur atom of cysteine bound in protein. (49, 56) Cell respiration is thus a multiple iron catalysis. As early as 1933, at a time when the concept of the respiratory chain had not yet been formulated, Warburg described the role of iron by the scheme [218]:

$$O_2 \rightarrow \overbrace{\text{ferro} \rightarrow \text{ferri}}^{1} \rightarrow \overbrace{\text{ferro} \rightarrow \text{ferri}}^{2} \rightarrow \overbrace{\text{ferro} \rightarrow \text{ferri}}^{3}$$

$$\overbrace{\text{ferro} \rightarrow \text{ferri}}^{4} \rightarrow \overbrace{\text{ferro} \rightarrow \text{ferri}}^{5} \rightarrow \ldots \text{substrate}$$

This scheme indicates that respiration is a catalysis in which a chain of iron compounds, arranged in series, act catalytically by the change in valency of the iron atoms.

The 'non-haem' iron catalysts differ in respect to their redox potential which lies between -0.5 and $+1.0$ volts.

Warburg's next major undertaking was the elucidation of the mechanism of action of the 'hydrogen-activating enzyme' in biological material. It had been known from the work of Thunberg, Wieland, and others, beginning in the first decade of this century, that biological material contains enzymes which catalyse the reduction of methylene blue (and similar synthetic dyes) or nitrophenols to colourless compounds by a variety of substances like succinate, malate, citrate, lactate, or glutamate. The common feature of these substances is their ability to donate hydrogen atoms and undergo 'dehydrogenation'. Such experiments indicated that O_2, the physiological hydrogen acceptor in the process of combustion, can be replaced by other acceptors. During the 1920s it was intensely argued whether the catalytic activation of oxygen, as studied by Warburg, or the catalytic activation of the hydrogen atoms of the substrates, as studied by Thunberg and by Wieland, was the essential feature of biological oxidations. At the earlier stages of the discussion Warburg dismissed the concept of the activation of hydrogen because the experiments of Thunberg and others on which it was based involved non-biological oxidizing agents such as methylene blue or nitrophenols. He scathingly stated that the oxidation of organic material by derivatives of nitric acid or synthetic dyes was of no biological interest. But he did not properly appreciate that although methylene blue is not a naturally occurring compound, the reactions which Thunberg had studied were enzyme-catalysed processes. Eventually, however, Warburg became impressed by the enzymic nature of the methylene blue reduction when E. S. G. Barron demonstrated some methylene blue experiments on red cells to him personally in 1929 during a visit to the Johns Hopkins Medical School at Baltimore. A year earlier Barron and Harrop (5) had observed that addition of methylene blue to rabbit erythrocytes promotes an oxidation of glucose by molecular oxygen to CO_2 and pyruvate. In this system methylene blue was reduced enzymically and re-oxidized non-enzymically by O_2. Warburg was intrigued by this phenomenon and after his return to Berlin he began an investigation into the chemistry of the action of methylene blue. Initial experiments showed that O_2 was taken up on addition of methylene blue also after cytolysis of the red cells, provided that glucose was replaced by glucose 6-phosphate. Thus he obtained the oxidizing process in a homogeneous solution, and this made it possible to investigate the problem by the standard methods of chemistry. Dialysis of the solution indicated that the hydrogen-transferring mechanism involved a high molecular component—an enzyme—and a low molecular heat stable component—a co-enzyme. Next he discovered that yeast extracts showed a much greater activity under the same conditions and the subsequent work was therefore carried out on this material. The protein component was found to consist of at least two fractions. One of these was yellow and was later referred to as the 'yellow enzyme'. The second was colourless. On shaking with aqueous methanol the yellow enzyme was denatured and the coloured component went into solution. Warburg crystallized a derivative of the yellow component, luminoflavin, which Stern and Holiday (66) subsequently identified as a methylated alloxazine. Luminoflavin arises from the coloured component of the yellow enzyme by illumination in alkaline solution. Riboflavin (a substance which R. Kuhn first isolated from milk in 1933) on illumination forms the same luminoflavin. The yellow enzyme, as Theorell showed in 1934, consists of a specific protein and a phosphorylated riboflavin, the latter being the prosthetic group which undergoes reversible hydrogenation and dehydrogenation, according to the scheme

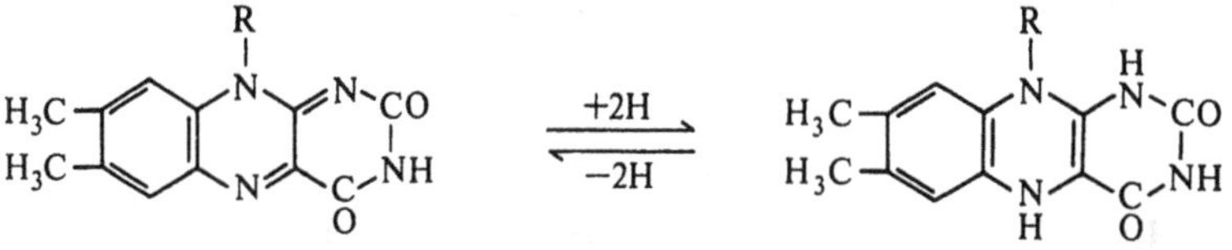

(oxidized flavin) (reduced flavin)

where R is ribitolphosphate. The ring system is that of an isoalloxazine. A variety of yellow enzymes, now called flavoproteins, were discovered, all containing riboflavin phosphate, and capable of acting as hydrogen carriers. In some cases the prosthetic group is the same as in Warburg's first yellow enzyme, namely a riboflavin phosphate or, as Warburg called it, 'alloxazine nucleotide' of the general structure:

Phosphate—ribitol—dimethylisoalloxazine

but in the majority of cases the prosthetic group is a 'flavine adenine dinucleotide':

Phosphate—ribitol—dimethylisoalloxazine
|
Phosphate—ribitol—adenine

This dinucleotide was also discovered by Warburg. He identified it when purifying D-amino acid oxidase. Later it was established as the prosthetic group of other oxidizing enzymes, including glucose oxidase, xanthine oxidase, and acyl CoA dehydrogenase. In the latter cases the flavoproteins accept hydrogen atoms directly from organic substrates, but this does not apply to the red-cell system where Warburg found the first yellow enzyme. As already mentioned this requires a coenzyme, the nature of which Warburg proceeded to investigate, a line of work which was to lead to another outstanding achievement.

Adapting classical organic chemical methods of isolation, Warburg obtained a pure coenzyme in 1934. [225, 235, 243, 255] It contained nitrogen and phosphorus, and on hydrolysis formed phosphate, pentose, adenine, and a substance not hitherto known to occur in coenzyme fractions. This was identified as nicotinamide:

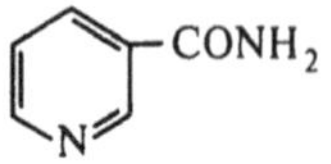

Hugo Theorell, who was an eye-witness of this discovery, gives the following description (67) of the circumstances:

In December 1933 he showed me the first crystals of nicotinic acid amide as picrolonate. Nobody knew at that time what it was; that had to be found out. When I told Warburg I might go home to Stockholm for Christmas he hesitated, because there were living dangerous people like Hans von Euler and Karl Myrbäck who were on the same track. He finally agreed, but said: 'I am going to kill you if you say the word picrolonic acid in Stockholm.' That was easily promised.

The yield of the substance was poor, to say the least: a few milligrams from each batch of two hundred litres of horse blood. In order to find out the constitution by conventional means Warburg calculated that he would need so much blood that it would involve killing all the horses in Germany. Being a great lover of horses as a cavalry officer from World War I he did not like this idea, so something else had to be done. Fortunately, they had the elementary analysis, melting point and the molecular weight. Now a friend of Warburg's, Walther Schöller, who was the head of the Schering Kahlbaum Company Laboratory, made the simplest trick in the world: he looked into 'Beilstein' for substances with the same composition and melting point and within no time he said:

'Well, this is nicotinic acid amide, synthesized by Mr so-and-so in 1878 or something like that.'

Warburg's comment was laconic as usual: 'Yesterday we could not buy it for any money in the world, today we can buy it for two marks a pound.'

It is not often that such things happen but let us agree that the few of us who have ever witnessed such explosional progress will never forget it. These are the golden moments in the life of a scientist when a sudden result comes out so clearly that the experiment does not need to be repeated.

Nicotinamide eventually turned out to be the catalytically active group of the coenzyme which, by undergoing hydrogenation and dehydrogenation, transfers hydrogen atoms. The same substance was subsequently found to be present in the coenzymes of numerous other oxido-reductions, especially those of fermentations. By now well over 150 different dehydrogenases are known where nicotinamide plays a key role as a hydrogen carrier.

The discovery of nicotinamide and the elucidation of its mode of action was a monumental achievement. Many had tried before Warburg to isolate and identify the active principle in the coenzyme fraction of oxido-reductions, especially of fermentations. The existence of coenzymes had been known since the pioneer work of Harden in 1906 but its function in yeast fermentation remained entirely obscure. The presence of adenine, pentose, and phosphate had already been established by von Euler in impure fractions of the coenzymes of alcoholic fermentations; however as long as the substances were impure, the significance of the components isolated from it remained uncertain. Warburg made full use of the facilities for large-scale operations which he had installed in his new institute in 1931. To illustrate the scale of his operation: the starting material for the isolation of the coenzyme was 100 litres of washed horse erythrocytes.[243] These were lysed with 200 litres of water and at once treated with 500 litres of acetone. This yielded 4.8 g of 'coenzyme'. Micro-tests monitored the coenzyme content at each stage of the purification procedure.

The first pure coenzyme to be isolated by Warburg contained one molecule nicotinamide, one molecule adenine, two molecules ribitol, and three molecules phosphate. It proved to be a dinucleotide of the general structure

```
              phosphate
                  |
   phosphate—ribitol—adenine
        |
phosphate—ribitol—nicotinamide
```

Soon afterwards Warburg isolated, also from red cells, a second coenzyme which differed from the first by containing two phosphates only. This was the coenzyme of glycolysis and of alcoholic fermentation. Since the pyridine ring is the characteristic feature of these dinucleotides, Warburg named them pyridine dinucleotides and distinguished between the diphospho- and triphospho-pyridine nucleotides (DPN and TPN). This nomenclature was in use for over thirty years until the International Nomenclature Committee replaced it by the terms nicotinamide adenine dinucleotide and nicotinamide adenine dinucleotide phosphate (NAD and NADP).

The identification of nicotinamide in the coenzyme fraction did not at once explain how the coenzymes act, but the availability of pure substances made it possible to study their reactivity under clear-cut conditions. Glucose 6-phosphate and TPN, together with a specific protein isolated from red cells led to a prompt interaction according to the equation

pyridine nucleotide + glucose 6-phosphate + H_2O
→ dihydropropyridine nucleotide + phosphogluconic acid.

Thus hydrogen had been transferred from glucose 6-phosphate to the pyridine nucleotide. The dihydropyridine nucleotide was not auto-oxidizable but on addition of the yellow enzyme the following reaction occurred

dihydropyridine nucleotide + yellow enzyme
→ pyridine nucleotide + reduced enzyme.

By these two reactions hydrogen is transferred from glucose 6-phosphate to the yellow enzyme, with the pyridine nucleotide acting as a hydrogen carrier. Warburg suspected that this carrier function is brought about by a reversible hydrogenation of one of the double bonds of the pyridine ring although partial reversible hydrogenations of this type were unknown to organic chemists at that time. On the basis of model-experiments, carried out jointly with P. Karrer [251] on N_1-substituted nicotinic acid derivatives, the conclusion was reached that the hydrogenation occurs at carbon atom 1 of the ring. Some 18 years later Pullman, San Pietro, and Colowick (59), using deuterium, established that in fact carbon atom 4 is the site of hydrogenation.

In the study of the reduction of pyridine nucleotides it was of great importance that on the reduction of the pyridine nucleotide an absorption band of 340 nm arises which provides a convenient spectrophotometric test for the reduction of pyridine nucleotides. The discovery of this absorption band was the basis of a large proportion of the subsequent work on the pyridine nucleotides, especially in the measurement of enzyme activities and the determination of minute quantities of metabolites. Warburg's collaborators, Negelein and Haas, made use of it when they examined the role of the specific proteins—the dehydrogenases—in the transfer of hydrogen. This work led to the conclusion that the pyridine nucleotides combine with the proteins and that the high specificity of the dehydrogenases resides in the protein part. Today this is taken for granted, but this was not so in the early 1930s, as Willstätter's statement of 1926 (72) illustrates: 'Enzymes are neither proteins, nor carbohydrates, nor do they belong to any of the known large groups of complex organic compounds.'

Thus Warburg's discoveries of the yellow enzyme and the pyridine nucleotides clarified key aspects of the mechanism of action of dehydrogenases.

Crystallization of enzymes of fermentation

By the early 1930s, thanks to the work of Harden, Neuberg, Meyerhof, Embden, the Coris, Parnas, Needham, and Lohmann, the enzymes of the intermediary stages of lactic and alcoholic fermentations had been identified and their reactions had been formulated, but not a single one of the enzymes had been obtained in a pure crystallized form. Since the ultimate analysis of the nature of enzyme action depends on the availability of pure substances, the purification of enzymes is of crucial importance. Only with pure substances can the stoichiometric interactions between enzymes, substrates, and coenzymes be analysed.

Warburg, together with his collaborators, Christian, Kubowitz, Negelein, and Bücher, tackled enzyme purification with new procedures and was the first to succeed in crystallizing enzymes connected with fermentations—no less than 9. In the now accepted nomenclature they were lactic dehydrogenase, enolase, aldolase (from muscle and from yeast), glyceraldehydephosphate, dehydrogenase, 3-phosphoglycerate kinase, alcohol dehydrogenase, pyruvate kinase, α-glycerophosphate dehydrogenase, and triose phosphate isomerase. Others, e.g. glucose 6-phosphate dehydrogenase, were highly purified.

The starting points for this striking progress were the new tests of enzyme activity. Prior to Warburg, the quantitative assay of enzyme activity, essential in the process of purification because it indicates the enzyme content of each fraction, was usually very laborious. It involved the chemical determination of substrate or product changes and this often required hours rather than minutes and also consumed considerable amounts of valuable material. The new optical tests could be carried out in minutes on micromole quantities. Since in the key steps of fermentation, reduced pyridine nucleotide is either formed or removed, these steps can be tested quantitatively by measuring changes in the absorption at 340 nm. Furthermore steps not involving pyridine nucleotides can be coupled to such a reaction and thereby become amenable to the optical test. A second absorption band which Warburg found useful is the enol band of phosphopyruvate at 240 nm. He introduced many other methodological details such as the use of protamine for the removal of nucleic acid from protein solutions, a technique now commonly applied.

A consequence of the work on enzyme solution was the discovery of a new key intermediate of fermentations: 1, 3-diphosphoglycerate, a discovery only possible after the purification of the glyceraldehydephosphate dehydrogenase.[288] The purification of the enzymes also led to the clarification of the inhibition of fermentation by fluoride. Lohmann and Meyerhof had established that fluoride specifically inhibits enolase. Using the pure enzyme Warburg showed that the inhibitory substance is a Mg-fluorophosphate which combines with the enzymes.[293]

The crystallization of the enzymes of fermentation and of the coenzymes not only elucidated finer details of the intermediary steps of fermentations but also had far-reaching practical consequences. It provided powerful microanalytical tools, the crystalline enzymes being extraordinarily specific and highly sensitive agents for the analysis of cell constituents. Such reagents proved decisive in many other fields of research, as well as clinical biochemistry.

Pentosephosphate cycle

Warburg's work also laid the foundation for the discovery of the pentosephosphate cycle, that is, the reactions which supply pentosephosphate for the synthesis of nucleotides, nucleic acids, and reduced triphosphonucleotides (NADPH). He discovered the first step of the pentosephosphate cycle, the formation of phosphogluconic acid from glucose 6-phosphate [243], a dehydrogenation where NADP is the hydrogen acceptor. This reduced triphosphopyridinnucleotide is used in many syntheses and in detoxication processes in the liver.

Photosynthesis

Warburg's general interest in biological energy transformations led him in 1914 to start a study of photosynthesis—the most important of all energy-transforming mechanisms in living matter, because all life depends on it. One particular aspect of this field fascinated him: the thermodynamic efficiency of photosynthesis. Connected with this was the problem of the *nature* of the process in which light participates, the 'photochemical reaction'.

In 1912 Einstein had formulated the law of photochemical equivalence according to which a photochemical reaction involves primarily the absorption of 1 quantum per 'photolyte', i.e. the substance becoming reactive after absorption of light. Soon afterwards, Emil Warburg provided the first experimental evidence of the validity of this law in studies on the photochemical fission of HBr and HI. Otto's close personal contact with these developments provided the basis of his approach to the study of photosynthesis. It was evident to him that this problem could only be tackled by new methods of measurement. The necessary physical methods for quantitative determinations of light intensities—bolometry—had been worked out in his father's laboratory where he had learned these techniques. Light intensity measurements had to be correlated to measurements of the rate of photosynthesis and when Warburg entered the field reliable quantitative methods for measuring the rates under a variety of conditions were very limited. He introduced several decisive innovations. Firstly, he modified the manometric technique (which he had already used for the measurement of cell respiration) to the study of photosynthesis. Secondly, he adopted suspensions of a unicellular alga—*Chlorella*—as experimental material, having in earlier work found suspensions of other living cells—erythrocytes, yeast, staphylococci—to be very valuable material for quantitative work. Investigators before him had used leaves, which do not lend themselves readily to quantitative experiments. Suspensions of *Chlorella*, which can be grown under precisely controlled conditions and can be handled like solutions, were an ideal material for quantitative work on photosynthesis, and have been favoured ever since. Thirdly, he developed the method of intermittent illumination which proved essential in establishing conditions where the 'light reaction' (or reactions) rather than the 'dark reactions' of photosynthesis are rate-limiting. Fourthly, he introduced bicarbonate–carbonate buffers which solved the problem of controlling the CO_2 supply at low CO_2 concentrations. Fifthly, he developed the use of inhibitors and discovered that the 'photochemical reaction' is particularly sensitive to narcotics while the dark reactions, like respiration, are sensitive to cyanide.

One of Warburg's primary aims was to establish the validity of Einstein's law of the photochemical equivalence for photosynthesis. Using *Chlorella* cells, Warburg found that four light quanta were necessary to produce one molecule of oxygen (corresponding to the assimilation of one molecule of CO_2). In red light at 660 nm, which has a molar quantum energy of 43 kcal, a requirement of four quanta gave an efficiency of about 65 per cent, based on the requirement of about 115 kcal for the reduction of 1 mole of CO_2 to the level of carbohydrate (eq 1):

$$CO_2 + H_2O \xrightarrow{\text{light}} (CH_2O) + O_2 \qquad (1)$$

The finding that about 65 per cent of the absorbed radiant energy can be transformed into chemical energy was remarkable not only for its high efficiency but also for the requirement of four collaborative quantum absorption acts to bring about the evolution of one molecule of O_2—a situation in conflict with Einstein's law which postulates that the primary process in any photochemical reaction involves the absorption of a single quantum. It was on these grounds that as early as 1926 Henri (31) had suggested that oxygen evolution by whole cells (i.e. complete photosynthesis) is unsuitable for measurements of the quantum efficiency of the primary photochemical act in photosynthesis. Warburg responded by postulating that CO_2 molecules remain adsorbed on the chloroplast surface until successive step-by-step reductions by light-activated chlorophyll convert them to glucose and liberate oxygen. By invoking such a mechanism, Warburg continued to regard the oxygen evolution that accompanies CO_2 assimilation as a valid measurement of primary photochemical reactions.

In the later years many investigators who used Warburg's methods obtained much higher values for the quantum requirements of photosynthesis: ten or more quanta per molecule of oxygen evolved. These results did not shake Warburg's conviction about the unique efficiency of photosynthesis. During his last 20 years he vigorously reaffirmed his earlier findings and, together with D. Burk [325, 326, 329], advanced them to the point of dividing the process of photosynthetic energy conversion into two parts: (a) a one-quantum light reaction which liberates oxygen and converts a bound species of CO_2 into carbohydrate; and (b) dark oxidative reactions which provide the rest of the total energy needed for CO_2 assimilation. He viewed the one-quantum light reaction as the long-sought primary reaction that conforms to Einstein's law of photochemical equivalence. In his last major paper on photosynthesis, published in 1969 [497], Warburg represented this one-quantum light reaction as:

$$1 \text{ photolyte} + 1\, Nh\nu = 1HCOH + 1O_2 + 1 \text{ chlorophyll} \qquad (2)$$

where photolyte is defined as 'one molecule of carbonic acid combined with one molecule of chlorophyll'. Warburg's concept envisaged that reaction (2) supplied one-third of the total energy required for CO_2 assimilation, with the remaining two-thirds of the energy (used in the formation of the photolyte) being provided by reaction (3), i.e. the dark oxidation of two-thirds of the carbohydrate product (HCOH) formed in reaction (2):

$$\tfrac{2}{3}HCOH + \tfrac{2}{3}O_2 = \tfrac{2}{3}H_2CO_3 + 75 \text{ kcal.} \qquad (3)$$

Thus the net carbohydrate product per quantum was 1/3 H_2CO, corresponding to a requirement of three quanta per molecule of CO_2 (reaction 4) and signifying an energy conversion efficiency of about 90 per cent:

$$1H_2CO_3 + 3Nh\nu = 1H_2CO + 1O_2. \qquad (4)$$

Warburg maintained that this extraordinary efficiency of photosynthesis occurs only under optimal conditions (which he described) when all of the chlorophyll is present as the photolyte, i.e. the chlorophyll–carbonic acid complex. Lower efficiences of photosynthesis, he reasoned, result from light absorption by a portion of chlorophyll that has not been converted to photolyte and is therefore photochemically inactive.

However, these concepts of the bioenergetics of photosynthesis have not gained general acceptance. Biochemists could not accept a mechanism for a one-step transformation of a compound at the level of carbonic acid to carbohydrate. Moreover, more recent research has amassed convincing evidence in favour of the reductive pentose phosphate cycle as the pathway of photosynthetic carbohydrate synthesis from CO_2. Furthermore, evidence has been accumulated in recent years that one-quantum light reactions in photosynthesis are concerned not with the terminal events of oxygen evolution and CO_2 assimilation but rather with intermediate light-induced electron transfer steps.(2, 3, 34)

The great majority of investigators agree with van Niel (55) that the photochemical reaction is a fission of water and not a direct reduction of CO_2. Water is taken to be split according to the reaction

$$H_2O + A \xrightarrow{\text{light}} \tfrac{1}{2}O_2 + AH_2$$

where A is a hydrogen acceptor. Direct experimental support for this concept came from experiments by R. Hill in 1939, who showed that isolated chloroplasts on illumination can form O_2 provided that a hydrogen or electron acceptor such as Fe^{3+}, e.g. ferric oxalate, is present. In 1944, Warburg and Lüttgens [315] found that quinone is a more effective hydrogen acceptor than Fe^{3+}. In 1960 Warburg showed that the Hill reactions are accelerated by catalytic amounts of CO_2 [427], and on the basis of this finding, Warburg remained convinced that the light reaction is a photolysis of CO_2 according to reactions (1) and (2) and chose to ignore the powerful argument that photosynthetic bacteria, which can also convert CO_2 to carbohydrate in light, do not produce O_2. Of course they require an electron acceptor in place of O_2, but sulphur compounds and H^+ can function as such. The primary electron acceptor in both cases is probably ferredoxin.

During the last years of his life Warburg talked several times about ways and means of bridging the gaps between his own views and those of others—without publishing the outcome. He visualized a conciliation of the divergent views on the basis of the fact that the Hill reaction depends on catalytic amounts of CO_2. This, he thought, may suggest that CO_2 participates in the Hill reaction as a catalyst, and as the Hill reaction is an oxido-reduction, the role of CO_2 may be that of an electron carrier. He had already formulated such a role in 1968:

$$H_2CO_3 + \text{light} \rightarrow HCOH + O_2$$
$$2NADP + HCOH + 2H_2O \rightarrow 2NADPH_2 + H_2CO_3$$

$$\text{sum: } 2NADP + 2H_2O + \text{light} \rightarrow O_2 + 2NADPH_2.$$

According to this mechanism carbonic acid would undergo reversible reduction and oxidation, and be regenerated after each catalytic cycle; it would not be converted to carbohydrate. This leaves room for the Calvin cycle (or analogous sequences).

Thus some aspects of the controversies may well resolve themselves on the grounds of the well-established fact that CO_2 is involved in photosynthesis in at least two entirely different ways, in a photochemical reaction (where it is a catalyst and may well react in the form of Warburg's postulated photolyte) and in the dark reactions of the Calvin cycle.

Meanwhile, Warburg's views on two interlinked aspects of photosynthesis—the energetic efficiency and the nature of the photochemical reaction—remain controversial. Nevertheless, despite what many would consider aberrations of judgement there is a wide agreement among his critics that Warburg's pioneering contributions to the experimental progress of the subject have been monumental.

Warburg's contributions to photosynthesis were not limited to the energetics of the process. Working with isolated chloroplasts he discovered in 1946 that chloride is an essential co-factor in the Hill reactions.[315] At that time it was not yet known that chloride is essential for plant growth. This is perhaps the only case in the field of plant nutrition where the site of action of a plant nutrient was established before it was known to be an essential nutrient. The essentiality of chloride for plant growth was established several years later.(8, 51) The precise role of chloride in the Hill reactions remains one of the unanswered questions in photosynthesis.[11]

There is evidence suggesting that chloride generally, not only chloroplasts, plays a role in the permeability of membranes.(4, 11, 35, 62)

Nitrate reduction in green plants

In addition to his work on photosynthesis Warburg made another important contribution to plant biochemistry through his studies of the mechanisms of the reduction of nitrate to ammonia.[45] Nitrate is reduced readily in the dark by green plants and by non-photosynthetic organisms, e.g. bacteria, but light was known to accelerate the process in plants. Before Warburg, quantitative information on the subject was limited because it was not possible to measure the rate of reduction accurately under controlled conditions, mainly because oxidizing processes very much predominated in cells which were capable of reducing nitrate. Warburg devised a simple technique which increased the rate of nitrate reduction to such an extent that it became the predominant process. He suspended *Chlorella* in mixtures of nitrate and nitric acid (10:1). On balance the following reaction took place:

$$HNO_3 + H_2O \rightarrow NH_3 + 2O_2. \qquad (5)$$

The evolution of oxygen or the formation of ammonia can be easily measured. Whilst the reaction occurs in the dark it is greatly accelerated by light. Warburg's investigations showed that the mechanism of this process is not adequately expressed by the balance equation. Nitrate reduction is interlinked with the metabolism of carbon and, on illumination, with photosynthesis. The first step is a dark reaction in which nitric acid interacts with organic substances to form ammonia and CO_2:

$$HNO_3 + H_2O + 2C \rightarrow NH_3 + 2CO_2. \qquad (6)$$

Another step is photosynthesis which regenerates organic carbon from CO_2:

$$2CO_2 \xrightarrow{\text{light}} 2C + O_2. \tag{7}$$

The balance of these two reactions is identical with the balance of nitrate assimilation (reaction 5). Since the carbon used in reaction (6) is regenerated in the light, small amounts of carbon can reduce unlimited amounts of nitrate in the light.

These papers appeared to answer the question of the mechanism by which light accelerates the reduction of nitrate—though many years later the answer turned out to be incorrect. Nevertheless, here again Warburg had laid the foundations for subsequent work in other laboratories. His findings certainly had justified the assumption that the reduction of nitrate in green plants is not directly connected with photochemical processes, and this concept was generally accepted for several decades. Eventually the clarification of the component reactions of photosynthesis revealed that the reduction of nitrate in light is linked to photosynthesis—without participation of carbohydrate or carbon. The energy of light reduces NADP, flavinucleotides, and ferredoxin, and these reduced coenzymes reduce nitrate to nitrite, and nitrite to ammonia.(7, 10, 47, 48)

Ferredoxin

Warburg was also involved in the discovery of the electron carrier ferredoxin mentioned in the preceding section. This substance is distinguished by its strongly negative potential, close to that of hydrogen gas: ($E_m = -0.420$ V at pH 7.0). Its discovery has several roots. The first experiments dealing with this substance were carried out in Warburg's laboratory by Kempner and Kubowitz in 1933 who discovered that the butyric acid fermentation in *Clostridium butyricum* is reversibly inhibited by carbon monoxide and that this inhibition is reversed by light.[217] This light effect made it possible to measure the 'action spectrum' of the catalyst: it was not that of iron porphyrin but that of a non-haem iron protein. Independent of this work was the discovery in 1952 by Davenport, Hill, and Whatley (15) of a factor, which eventually proved to be ferredoxin, catalysing the reduction of methaemoglobin by chloroplasts. Also independent was the work by Mortenson, Valentine and Carnahan (54) in 1962 which led to the isolation of an iron-containing enzyme from *Clostridium*. This contained 2.5 per cent iron and was named ferredoxin. At the same time Gewitz and Völker [453] in Warburg's laboratory isolated a 'red enzyme' from *Chlorella* which contained 1 per cent iron and produced one molecule of H_2S per atom of iron on addition of dilute sulphuric acid. Such a release of H_2S was later found to be the property of ferredoxins from all other sources. Gewitz and Völker also established an inhibition of photosynthesis by carbon monoxide and noted that the 'action spectrum' of the inhibited catalyst was that of a non-haem iron protein very similar to that found by Kempner and Kubowitz in *Clostridium butyricum*.

Clinical biochemistry

In 1927 Warburg [108] designed a method of determining trace quantities (10^{-4}mg) of Fe and Cu based on the catalytic effects of these ions on the oxidation of cysteine by molecular oxygen, and with these methods he established the basic facts concerning the occurrence in biological material of free or 'loosely bound' iron and copper (i.e. Fe and Cu which readily chelate with cysteine). He discovered that Fe and Cu are regularly present in human and animal blood plasma and that their concentrations under normal conditions are as constant as are other blood constituents. He established the normal range which for both Cu and Fe is of the order of 1 mg/litre and found that the concentration of copper is raised in pregnancy, infections, and iron-deficiency anaemias and that the Fe content falls in iron-deficiency anaemias.[116, 118] This work was the starting point for the clinical application of Fe and Cu determination in human serum or plasma, a matter of diagnostic importance especially in Wilson's disease where the Cu content is greatly decreased.

About five years later, Ludwig Heilmeyer began to investigate the biochemistry and pathology of iron (29) and later of copper (30) in blood plasma. He introduced the spectrophotometrical methods for the quantitative determination based on the formation of coloured compound with metal-chelating agents. Since then the estimation of iron and copper in plasma has become clinical routine. The determination of iron is valuable in the differential diagnosis of anaemias. Plasma copper is decreased in Wilson's disease, while the copper content of the tissues is raised. It is now known that Warburg's 'loosely bound' metals in blood plasma are present as complexes with specific proteins—ferritin and coeruloplasmin. These metal proteins are transport and storage forms of the metals.

Warburg was the first to detect enzymes in blood plasma which normally function only within living cells. These were, in the first instance, lactate dehydrogenase and aldolase. As they have no function to perform in the plasma, Warburg reasoned that they must be derived by a leakage from cells, or by a disintegration of cells. The detection of the enzymes was possible because of the highly sensitive optical tests. Warburg also found in 1943 that the aldolase content of plasma can rise in pathological states, e.g. in tumour-bearing rats. Years later, enzyme assays in blood plasma became established methods of clinical biochemistry. In his book *Enzymes in blood plasma*, B. Hess (33), wrote in 1963:

> A decisive stimulus to clinical enzymology was given by Otto Warburg in 1935 as a result of the discovery of optical means in the determination of enzyme activity, and again in 1943 by the demonstration of enzymes of glycolysis in the serum of normal and tumour-bearing rats. As early as 1943, Warburg and Christian opened up the entire biochemical and biological problem of the phenomenon of serum enzymes.
>
> Their results were confirmed in 1949 by Sibley and Lehninger (64) who carried out systematic investigations on animal and human serum. Later, after 1953, several other investigators discovered the presence of enzymes in mammalian serum and noted that their amounts vary under pathological conditions. This information attracted much attention and soon the assay of serum enzyme activities became a routine diagnostic test.

Other practical applications

Some of Warburg's work, although aimed solely at gaining knowledge of the laws of nature, had other far-reaching practical and economic consequences. His discovery of nicotinamide as a cell constituent in 1935 led two years later, in the hands of

Elvehjem (18), to its identification as the anti-pellagra vitamin in liver extracts and to the possibility of preventing and curing pellagra. In 1945 Chorine, when studying the effects of vitamins on mycobacteria, discovered the powerful inhibitory effects of nicotinamide on the mycobacteria responsible for leprosy and tuberculosis.(12) This was the starting point for systematic investigations into the anti-tubercular effects of substances related to nicotinamide and, in the laboratories of the Hoffmann–LaRoche Company, led to the discovery that isonicotinic acid hydrazide ('isoniazid') is even more effective and has fewer side-effects than nicotinamide. This drug is now widely used in the control of tuberculosis.[12]

Warburg's research on the crystallization of enzymes and on measuring their activity by spectrophotometric methods was the basis of a major industrial development—the manufacture and marketing of pure enzymes and of their co-factors and substrates. Crude enzyme preparations, such as urease, pepsin, and pancreatin had been marketed for many decades, some for medical purposes, others as useful chemical agents, but their scope was limited. The manufacture of crystalline enzymes and the development of spectrophotometric methods provided much more specific—in fact, uniquely specific—and much more sensitive tools. The replacement by fluorometry of spectrophotometry further increased sensitivity by several orders of magnitude.

Dr. H. U. Bergmeyer who, beginning in the late-1940s, built up the enzyme manufacture of the Boehringer Mannheim Company states that without Warburg the enzyme branch of Boehringer Mannheim would not exist: 'We consider ourselves as the executors of Warburg's pioneer work.' A little earlier, in 1947, Dr. Charles C. Worthington had started to initiate other aspects of industrial enzyme production in the United States. He had been trained in the laboratory of Dr. M. Kunitz at the Rockefeller Institute in Princeton where hydrolytic enzymes were first crystallized on a major scale, and began with the manufacture of trypsin, chymotrypsin, ribonuclease, and deoxyribonuclease. In 1954 the Seravac Company of South Africa began to manufacture some of the enzymes which Warburg had first purified and crystallized (alcohol dehydrogenase, lactate dehydrogenase, malate dehydrogenase). A managing director of the company, Dr. B. R. Cookson, testifies that Warburg's work was vital to the company's enterprises.

Warburg had not, of course, set out to create a new branch of the chemical manufacturing industry. He needed the pure enzymes in order to study fundamental processes in living cells. He did not even think of suggesting to firms that they might manufacture enzymes. This was done by Bücher, who approached the Boehringer Mannheim Company. This account illustrates the frequent experience that first-rate fundamental research, sooner or later, leads to important practical applications.

The last decade

Right to the end of his life Warburg published experimental papers at regular intervals at the rate of about five per year, in the later years mainly on photosynthesis and on cancer. In recognition of his outstanding achievements the Max Planck Gesellschaft had waived the usual retirement regulations and fixed no age limit. Although he spoke occasionally of retirement it was his view that as long as he could personally control the work in the laboratory he should carry on, but he announced that he would retire as soon as he felt that he could no longer check every detail personally. To Warburg, that time never came. Before retiring he hoped, so he stated, to complete the work on the quantum physics of photosynthesis and to find the cure for cancer. This, he thought, should be possible.

As his bibliography shows, the papers of the last period still contain many valuable contributions to both fields of his choice. An example is the discovery of the involvement of CO_2 in the Hill reaction [427], made in 1960 when he was 77. But most of the work had no longer the pioneering quality of his first 45 years of research. While pursuing ideas based on the old concepts he lost contact with the main stream of progress. In both fields new aspects came to the forefront. Warburg's concern was almost exclusively with the energy aspects which at the time he began were indeed outstanding problems, and he limited his interests to the over-all balance of energy changes. He took no serious interest in the intermediary steps of energy-transformations in photosynthesis, or in the general biochemistry of growth and the mechanisms which control growth.

Although his drive, industry, and experimental skill remained, towards the end his ability to keep abreast of the wider aspects of his fields and to judge objectively the merits of his ideas and experimental observations slipped from his former high standards. If he lost contact with the newer developments it was perhaps in part due to his unwillingness to surround himself with independent fellow scientists and to listen to their points of view, their criticisms, their approaches, and their interests. His technical collaborators were a tremendous support to him but they could not act freely in a critical and independent role. So he forewent the benefits which all scientists derive from free and frequent contacts with their equals and their students, including the youngest among them.

Warburg was occupied with his work almost until the last day of his life. On 24 July 1970, he decided to stay at home because he did not feel well and on the next day he felt a pain in his leg, which had suffered a fracture of the neck of the femur in November 1968. The doctor diagnosed a deep vein thrombosis. Warburg stayed at home and continued to read and to write until 1 August when he felt rather weak. Late in the evening of that day a pulmonary embolus suddenly ended his life.

Formal recognition

The many honours bestowed upon Warburg[13] include, apart from the Nobel Prize, Foreign Membership of the Royal Society (1934), an Honorary Doctorate of Oxford University (1965), the German Order of Merit (restricted to 30 of Germany's most distinguished citizens), and the Freedom of the city of Berlin. Quite a few of the honours which he was offered he felt unable to accept, among them an Honorary Doctorate of Newcastle University, because he did not wish to spare the time to have them conferred personally, a usual condition of honorary doctorates.

The book *Nobel, the man and his Prizes*, published by the Nobel Foundation (46), which gives glimpses of the work of the Nobel committees, records that on no less than three occasions Warburg was considered worthy of a Nobel prize, and each time to recognize a different piece of work. In 1927 it was his work on the metabolism of cancer cells; the proposal was that the Prize should be divided with Fibiger but the Faculty preferred to give Fibiger the undivided Prize for his discovery of the Spiroptera carcinoma. Incidentally, the subsequent evaluation of Fibiger's work cast grave doubts on the wisdom of this decision. In 1931 Warburg was awarded the Prize for his discovery of the catalytic role of iron porphyrins in biological oxidations. In 1944 he was again found to deserve the honour for identification of the flavins and of nicotinamide as hydrogen carriers in biological oxidations, but Hitler's decree which forbade the acceptance of Nobel Prizes by German citizens intervened.

Warburg was of course not alone in deserving more than one Prize. Emil Fischer, Ernest Rutherford, Albert Einstein, Max Born, and others all had several achievements to their credit which deserved a Nobel Prize. Evidently there are scientists who maintain the highest standard of achievement over long stretches of their lives; whose Prizes are not merely a question of one lucky hit (as they may be in some cases) but an expression of sustained greatness in research.

Among the honours that Warburg received, he appreciated especially the honorary doctorate of Oxford University. The ceremony took place under the Chairmanship of the Chancellor of the University, Harold Macmillan (former Prime Minister) on 23 June 1965 on 'Encaenia' Day, the annual dedication festival to commemorate the Founders and Benefactors of the University. One of the eight honorary doctorates was the Prime Minister of the time, Harold Wilson. In the imposing Sheldonian Theatre, built by Christopher Wren in the late seventeenth century, each of the thousand seats was occupied for the occasion.

Much of the proceedings is still conducted in Latin, especially the conferment of the honorary degrees. The Speaker of the University—the Public Orator—Colin Hardie delivered the following speech when presenting Warburg to the Chancellor.

Necte tribus nodis ternos, Amarylli, colores!

haec apud Vergilium canit pharmaceutria. hospes hic noster, vir veritati deditus, tribus, ut ita dicam, scientiae coloribus praecipue imbutus tres potissimum nodos explicavit. chemiae doctrina ab Aemilio Fischer percepta, apud patrem suum didicit radiorum naturam, apud Ludolfum von Krehl artem medendi. mox suo marte quemadmodum cellae, quae vocantur, corporis animantis vi materiae acidae commutentur et quasi respirent—quam ad quaestionem idonea invenit ova echini—inter alia opus esse iis fermento quodam ferri particulis instructo repperit. deinde quomodo frondes solis lumine adiutae aëra inductum in nutrimenta convertant, herbam scrutando pusillam, quae chlorella appellatur, callide explicavit. ad cellas denique reversus, et in vitro, quod aiunt, et in corporibus murium experiendo, velut alter Hercules

malum immedicabile cancrum[14]

superare impigre conatus est.[15] si rationem operis poscitis, scitote eum ut opificem artificiosissimum minima haesitatione per machinas ingeniosas atque subtiles adsiduas adhibere periclitationes. praelegendi administrandique impatiens maximam longae atque uberis vitae partem, Americanis sane olim hospes gratissimus, in officina sua Berolini operatus, ne dictatoribus quidem graviter impeditus, indolis viribus acie ingeni maxima beneficia generi humano contulit. praesento vobis veteranum viridem, merito ornatum multis decoribus, praemio Nobeliano dignum bis existimatum. Othonem Henricum Warburg, biochemicorum antistitem primarium, ut admittatur honoris causa ad gradum Doctoris in Scientia.

Translation (by the Public Orator)

Tie, Amaryllis,[16] three knots of three hues!

So sings Virgil's sorceress. But our guest, a devotee of truth, imbued with the knowledge of three sciences, has untied in particular three knots. He learnt chemistry from Emil Fischer, radiation physics from his own father, medicine from Ludolf von Krehl. He then began his own research on the oxidation and respiration of cells—finding sea-urchin eggs suitable for his investigation—and discovered that this was made possible especially through an enzyme containing iron. His examination of the little plant Chlorella enabled him to analyse accurately the process of photosynthesis. Finally, by studying cells again, both *in vitro* and in mice, he made progress towards the conquest of cancer. His methods are those of an artisan who is an artist, ready, with the minimum of critical hesitation, to carry out innumerable experiments by means of the most ingenious and accurate apparatus. He does not like lecturing and administration; he has been a welcome visitor to America, but has spent the greater part of a long and fruitful life, not seriously disturbed even by dictators, in his Berlin laboratory, where his energy and intellect have made him the benefactor of mankind. I present to you a vigorous veteran, amongst other well-deserved honours twice adjudged worthy of the Nobel Prize, Otto Heinrich Warburg, doyen of biochemists, for admission to the honorary degree of D.Sc.

The Chancellor, Harold Macmillan, replied,

Doctor erudite, qui ingens artificis naturae opus tanta calliditate perscrutaris ut arcana cum mirabilia tum utilia reppereris, ego auctoritate mea et totius Universitatis admitto te ad gradum Doctoris in Scientia, honoris causa.

(Translation)

Learned doctor, you have studied the great work of nature the craftswoman and have revealed secrets that are both wonderful and beneficial; with my authority and that of the University as a whole, I admit you to the degree of Doctor of Science *honoris causa*.[17]

The rapid progress of science and the volume of material to be covered in textbooks and in teaching leaves little time for historical aspects. History has been crowded out and the new generation tends to be ignorant of earlier pioneers. Many of Warburg's discoveries are now incorporated in the textbooks, often without reference to the originator.

Advances made in biology and biochemistry after Warburg were, of course, no less epoch-making. Examples are the elucidation of the three-dimensional structure of proteins, the sequencing of the amino acids in proteins and of nucleotides in nucleic acids, the discovery of the double helix structure of DNA, the deciphering of the genetic code, the elucidation of the mechanism of protein synthesis. Other outstanding achievements can be found among the Nobel Laureates after 1960. Progress in the second part of the twentieth century has indeed been stunning and much more rapid than scientists would have dared to predict.

Great though these achievements have been, in most of them team work played a rather greater role than it did in the discoveries made by Warburg, and great though the individual merits of the leaders have been, I think none has covered such a wide range of discoveries as did Warburg.

3
PERSONALITY

General characteristics

Warburg's great achievements in research, extending over a period of some 60 years, are the outcome of the combination of an exceptional intellect with a powerful determination to devote his life to experimental science. Ingredients of his intellect were a penetrating intelligence, an original and imaginative approach to any situation, and independence from commonly-held beliefs, judgements, and prejudices. Like Pasteur, whom he admired greatly, he was impelled by a powerful urge. His involvement with science was the dominant emotion of his adult life, virtually subjugating all other emotions. He never married. As a young student in Freiburg he was for some time deeply in love with the daughter of one of his father's colleagues but he decided, as he did also on a later occasion, that marriage would be incompatible with his professional ambitions.

In his every-day life Warburg followed a very regular routine and paid a great deal of attention to health and fitness. For many years he rose at 5.30 a.m. and went riding from 7 until 8 (when there was daylight). By 8.00 a.m. he was in the laboratory and, except for a lunch break, remained at his bench until 6.00 p.m. or so, and woe betide fellow-workers who left the laboratory any earlier! He used to spend every August and September at his country house on the Isle of Rügen in the Baltic, where he worked, rode, and spent many hours writing his papers and reading. He took all his notebooks with him. In later years, after the partition of Germany in 1945, he went instead to the 'Wiesenhof', a country estate in the Hunsrück Hills, south of the Moselle valley, which belonged to Jakob Heiss, Warburg's devoted companion from 1919 to the end.

Horse-riding was Warburg's favourite physical recreation. He kept his own horses, at times three, and even during the Second World War he managed to keep two, thanks to the help of the officers of a Cavalry Regiment. In 1924 he fractured his pelvis in falling from his horse and for some time lay, helpless, in a wood. He maintained that it was entirely the fault of the horse, but thereafter he never rode unless Heiss was with him. In 1955 Heiss had to give up riding after an operation. Warburg then had a riding area set up in his own garden and kept only one horse, a grey Hanoverian mare. He rode regularly until he was 85 when another fall, this time from a ladder in his library, fractured a femur. The same accident put a stop to his sailing, a sport he had taken up with fanatic zeal in 1952. For 16 years he had sailed on the river Havel almost every Sunday, and he hoped to be able to take it up again when his fracture had mended. Only in May 1970 did he reluctantly part with his boat. He was fond of the countryside around Berlin with its many lakes and woods and enjoyed walking there, especially on Sundays in winter. It is a measure of the respect in which he was held that, by the special and most exceptional courtesy of the Government of the German Democratic Republic, he had a pass which allowed him to roam at any time outside the city—this had been mediated on his behalf by Professor Manfred von Ardenne.

I owe to Karlfried Gawehn, Warburg's collaborator from 1950–64 the following description of the working style in Warburg's laboratory during those years.

The official working hours in the 1950s were from 9 a.m. to 6 p.m. Monday to Friday, with a mid-day break from 1 to 2 p.m.; on Saturdays we worked from 9 a.m. to 1 p.m. Warburg usually arrived shortly before 9 a.m. and it was a matter of course that everyone else was there by 8.45 a.m. at the latest. We usually extended our luncheon break a little because Warburg was normally away for 2½ hours. He left with Heiss at about 1 p.m. for his home (1 minute away) and after lunch would take a brief nap and not come back before 3.30 p.m. Nevertheless, our working day was at least 8 hours because Warburg did not leave the Institute before 6.30 p.m. Here again it was accepted that we did not leave before him, if only because he often required some of our latest results for his evening work at home. His working week was regularly about 60 hours.

Warburg failed to appreciate that people might occasionally need an hour or two's leave of absence to deal with personal matters—for example to visit a doctor or bank or child's school. When asked for permission he would say, 'You have enough time in the evening to attend to such things and besides, you have at your disposal 7 to 8 weeks' leave each year.' In the early 1960s the Directors of the Max Planck Institutes were discussing a revision of working hours. Warburg's comments were characteristically humorous-cum-aggressive: 'Can you imagine,' he said, 'those people' (meaning his fellow-directors) 'no longer wish to work on Saturdays? When I asked them why they considered it no longer necessary, they said that employees needed time for shopping. So I told them that perhaps they should also give their employees Friday off so that they could do still more shopping.' To me this answer sums up Warburg's attitude very well: for him there were no reasonable grounds, apart from death, for not working. Illnesses such as influenza were no excuse for absence.

Warburg himself adhered rigidly to this spartan style of working. When he was awarded a decoration by the Federal Government and was invited to the ceremonial investiture at the Town Hall, he telephoned the official: 'Send me my decoration by mail. My experiments allow me no time to leave my laboratory.'

At the end of the 1960s however, Warburg had to give up the fight and allow a five-day working week to be introduced.

The summer vacation was usually from the beginning of August until mid-September although Warburg himself preferred not to return before the middle of October in order to avoid celebrations and congratulations from colleagues on his birthday, 8 October. But there was no let-up in the laboratory between mid-September and mid-October. Work proceeded at full speed and results had to be reported to Warburg by telephone at least twice a day, in the morning at 9 a.m. and in the evening at 5 p.m.

When experiments were successful and the principles of his theories were not contradicted by experimental results, Warburg would be relaxed and would often reminisce about interesting happenings during the Hitler regime, during the War, and during his encounters with Russians and Americans after the War. He also talked about his private life. Mostly his talk was spiced with witty comments on politicians, officers, or his scientific opponents.

Warburg was an eccentric in his personal style, aristocratic in outlook (in the Greek sense of aristocracy, i.e. rule by the best), and, in consequence, he was masterful in his bearing. His mind was broadly based on a wide knowledge of history and literature. He was particularly fond of biographies. Much of his reading was in English.

Having had 4 years' war service in a crack cavalry regiment, he was much attracted by all that is best in soldierly conduct and attitudes: discipline and good manners, straightforwardness in thought and argument, the elimination of humbug and inefficiency. A military leader, he said, is relieved of his office if incompetent. In academic life, many senior people get away with manifest incompetence—technical and moral—whereas an army officer would not. A soldier, of whatever rank, is always at his post: academic colleagues, Warburg noted, took many liberties in absenting themselves from their primary duties. So he had rather more friends among soldiers than among university professors. Apart from his relationship with Jakob Heiss, none of his friendships gave the impression of being a close human relationship. On the other hand, he had great affection for his horses and dogs. He had a series of Great Danes—the last was called Norman, who weighed an exceptional 90 kg and slept on a couch in Warburg's bedroom where he received a piece of the best chocolate as a 'nightcap'.

Warburg's conception of personal honour and of what he would call 'gentlemanly conduct' are well expressed by comments he made on the fact that in the early-1950s Americans and Britons in Berlin had maintained social contacts with former National Socialist members despite warnings from Warburg about their affiliations.

In this context Warburg said that during his war service in the First World War, one of the officers invited some of his fellows to dinner. Among these guests were a Herr von Arnim and a Herr von Bismarck. When it came to the knowledge of Arnim that a Bismarck was to be in the company, he said to his host: 'Do you not know that since the year 1874 an Arnim has never sat down at table with a Bismarck? I must decline the invitation.' To Warburg, this conduct was 'correct'.

The story behind this odd behaviour is that Count von Arnim, German Ambassador in Paris in the early 1870s, incurred the displeasure of the Reichskanzler Otto von Bismarck. Allegedly Arnim had been active in favouring the re-establishment of the monarchy in France and because of this, Bismarck relieved him of his office. Soon afterwards, at the initiative of Bismarck, Arnim was summonsed on a charge of having appropriated important official documents. Arnim fled to Switzerland and in his absence was sentenced to five years' imprisonment for treason. There followed a very bitter private and public debate. The Arnim family and a great many other people considered the sentence as being without proper legal foundation. The Arnims therefore broke off all social connections with Bismarck, his children, and his children's children.(39) Warburg's comment was, 'This is the behaviour of gentlemen.'

Warburg was essentially a solitary man, self-sufficient, self-contained, never bored with his own company. His work, sports, reading, listening to his record collection, especially Beethoven and Chopin, occupied him to the full. Altogether he was a happy man right to the end. He found the greatest satisfaction in his successful creative work.

Warburg was most intent upon centring his life on matters which seemed to him worth while. He intensely disliked wasting time and carefully guarded himself against time-wasting activities and intruders who might waste his time. When he was working on cancer he was frequently badgered by aggressive reporters. When he told one that he had nothing new to report, he was asked to say something about future developments. Warburg declined this as unscientific. The reporter said that scientific accuracy did not matter; he was after something that sounded interesting. Newspapers, he said, were not necessarily concerned with truth but with entertaining the public. This was too much for Warburg, to whom the search for truth was everything, and he showed the reporter the door. This incident convinced him that he should never enter into any discussion with intruding journalists. Soon afterwards, another reporter appeared in the laboratory without warning, and said to the first person he encountered (which happened to be Warburg himself),

'May I speak to Professor Warburg?'

'Why do you want to see him?'

'I am a reporter and I wish to interview Professor Warburg about cancer research.'

'Sorry', replied Warburg, with straight face, 'He is dead.'

When he recounted the story to us later he could not suppress a smile of self-satisfaction.

Once a young lecturer in medicine wrote to Warburg, asking whether he could work in his laboratory. Warburg invited him for an interview, in the course of which he soon felt that the visitor would not be a suitable collaborator. Warburg asked why he was interested in his laboratory and the visitor replied that he was anxious to broaden his training in basic research. Warburg said swiftly, 'This I cannot provide. I am a specialist of the very worst type,' and ended the interview.

Another tedious visitor who declared a wish 'to lay all his cards on the table', was silenced by, 'Sorry! I never play cards. Good day!'

Nor was he afraid of annoying senior people. When he was taking leave of Simon Flexner who, as Director of the Rockefeller Institute,[18] had invited him to lecture at the Institute, Flexner said, 'Is there anything I can do for you?' Feeling this to be unduly patronizing, Warburg retorted, 'No, thank you. Is there anything I can do for you?'

These incidents are characteristic of Warburg's sense of humour. Throughout his life Warburg typed his papers and letters himself, and each paper was typed many times. The Institute employed secretarial help on a part-time basis only and this was for administrative, not scientific, work. He did not regard his typing as a waste of time. He answered letters promptly according to his belief that in a well-conducted office every letter is answered, if possible, on the day it is received. For many years he worked without a personal assistant, cleaning his glassware, preparing solutions, setting up routine tests, servicing his instruments, and looking after the animals.

He insisted upon being in every way master in his own laboratory. He alone decided on the appointment and dismissal of his collaborators. He once formulated his philosophy, quoting words of Goethe (26): 'I have been anvil long enough, now I want to be hammer.' As far as possible he resisted any interference from his employers, the Kaiser Wilhelm (later Max Planck) Gesellschaft. That he felt safe to do so is illustrated by the following incident which took place before the Second World War. One day he asked the office of the Director-General to allocate 10 000 marks for his research. He was told that the organization had no spare money; he should therefore write an application, which the Director-General would support, to the Notgemeinschaft der Deutschen Wissenschaft (an emergency fund made available by the government for the promotion of research). Warburg replied that he had no secretary; the Director-General should put a secretary at his disposal. This was agreed. Warburg handed the secretary a sheet of paper and told her to type, 'Top left: Dr. Otto Warburg; top right: the date; underneath in the middle: Application; underneath this: I require 10,000 (zehntausend) mark.' This Warburg signed. The 'application' was successful.

It was obviously Warburg's intention to make crystal-clear to the administration that, to a man of his reputation, their conduct was an unnecessary imposition and a bureaucratic waste of time.

His stature was short, but his bearing was upright. His speech was clear and to the point, without frills. So was his humour, as a postcard to his banker cousin, Eric Warburg, illustrates. Eric sent him a newspaper cutting about the award of a Nobel Prize to Hugo Theorell of Stockholm. Otto replied (Plate 16)

> Dear Eric,
> Many thanks for the newspaper cutting. Theorell has received the Prize for a paper which he carried out at my suggestion in my laboratory in 1934/5. This is now the third of my collaborators to receive the Nobel Prize—Meyerhof, Krebs, Theorell. Would not you also like to do some research with me?
>
> Cordial greetings, Otto.

There was a striking contrast between the fierce, aggressive, and self-righteous fighter whom his fellow scientists knew from the literature and conferences, and the Warburg in private social life or in the relaxed atmosphere of his laboratory. Socially he had an exceptional charm, when the company suited him. His conversation covered a wide range, was direct, original, penetrating, and often had a humorous and highly amusing flavour. The direct look from his bright blue eyes was both engaging and fascinating. Though very much a bachelor he could be particularly gentle, considerate, and understanding with women. An example is a letter to my wife (whom he liked very much). This letter was written in English and is quoted in full because it shows that Warburg could express himself well in English. The letter arrived out of the blue at Christmas time, five months after Warburg had visited us in Oxford:

Dear Lady Krebs, 20 December 1965

Perhaps you know, how enthusiastic we were about the visit of your Queen and the Duke. In memory of this event a medal was coined, that shows on one side the picture of the royal guests, on the other side the coat of arms of all the cities they visited.

Allow me to present you with this little medal in remembrance of another visit—the visit of a humble scientist in your beautiful home in the same summer.

With all good wishes for Christmas and looking forward to seeing you in June at Lindau.

Otto Warburg

People found it puzzling that in spite of his Jewish ancestry Warburg remained unharmed by the Nazis as far as his personal safety and his research work were concerned. One reason is that his mother's family was non-Jewish; another that Reichsmarshall Goering[19] (in accordance with his declaration 'I decide who is a Jew') arranged for a re-examination of Warburg's ancestry, and ruled that he was only a quarter-Jew. Perhaps the hope of a cancer cure from Warburg influenced this ruling. As a quarter-Jew, Warburg was not permitted to teach in a University, or to hold responsible administrative posts—which suited him very well—but he was not prevented from working in the seclusion of his laboratory. During the war years another important member of the Nazi hierarchy, Reichsleiter Bouhler, at one time chief of Hitler's Chancellery, protected Warburg on several occasions when he had been denounced for having spoken critically of the regime.

Warburg's willingness to let his Jewish blood be diluted in this way, and thus to make a pact with the Nazis, incensed colleagues outside Germany. He was not unaware of this, not least because he was a regular reader of *The Times* of London and quite well-informed about world opinion. Nor had he any illusions about the future of Nazi Germany. In 1938 he strongly advised Alexander von Muralt[20] not to accept an invitation to the Chair of Physiology at Berlin University, saying, 'Germany is heading for a great catastrophe.' He did not commit this to paper; nor would he say it until he was satisfied that the conversation could not be overheard.

But even in 1938 he was not afraid to poke satirical fun at Nazi officialdom. He had informed the organizers of the 16th

International Congress of Physiology to be held in Zurich that he intended to participate, but at the last moment he was instructed by the Nazis to 'cancel participation without giving reasons'. With some measure of courage he sent a telegram: 'Instructed to cancel participation without giving reasons. Warburg.' Professor A. V. Hill, one of the most distinguished British participants, was keen to have the telegram read out at a plenary session of the Congress but its President, Professor W. R. Hess, pleaded that it might provoke unpleasant international repercussions and that the organizers ought not to risk annoying the rulers of Germany. So Warburg's message was not read out in public: it was 'leaked out' and rapidly spread by word of mouth.

Warburg defended his compromise with the Third Reich on the grounds that he was, first and foremost, a research scientist, and one of very exceptional ability. He regarded it as his mission to serve humanity by advancing knowledge, especially in a broad-based fight against cancer. Writing in 1946 after the death of Hans Fischer, the undisputed master of porphyrine chemistry, Warburg said, 'The oft-quoted remark that no-one is irreplaceable is obviously not true in science.' (Ref.[506], p. 14.) Warburg thought of himself also as 'irreplaceable'. Emigration, he thought, would destroy his research potential because this depended greatly on collaboration with the team that he had built up over many years. If he left Germany the team would disappear. He once said, with reference to the emigration of Meyerhof: 'It is difficult enough to find a place for a worker, but it is impossible to find a place for a king.' Nevertheless, after the war, when the Russians had taken all his equipment and he saw no prospect of carrying out experiments in Germany, he did not hesitate to sack all his longstanding collaborators—Negelein, Kubowitz, Christian—on the grounds that they had been disloyal to him. During the Nazi period, he claimed, they had reported to the Gestapo his casual criticisms of the Third Reich. Whether they did so will never be known. I believe there is a grain of truth in Warburg's allegation but to me their arbitrary dismissal was unjustified and very cruel. Professor Schöller has told me that Warburg had indeed been denounced but that he, Schöller, through his connection with Reichsleiter Bouhler, had been able to protect Warburg.

Warburg's collaborators were dismissed on his instructions through the office of the Kaiser Wilhelm Gesellschaft by a formal letter, with no personal explanation from Warburg. Some years later he told me that they had not only been disloyal but had also become senile. When I looked rather baffled by this remark—the collaborators were 15–20 years younger than himself—he said, 'Different people age at different rates.' Late in 1949 at the age of 66 he began to assemble and train an entirely new team of technical collaborators.

After the war, Warburg sounded out the possibility of working in the United States but his approaches were very coolly received. His laboratory, undamaged by bombing, had been requisitioned for use as the headquarters of the American forces of occupation. In 1950 the building was eventually restored to him and re-equipped.

Outlook on research

Defining his outlook in research, Warburg wrote, 'A scientist must have the courage to attack the great unsolved problems of his time. Solutions usually have to be forced by carrying out innumerable experiments without much critical hesitation.' [467] His attitude was based on a supreme mastery of experimental resources and technical skill: it enabled him to devise new experimental tools and thus to approach problems in a novel manner. He could penetrate areas which it had not been possible to explore before.

In a letter to D. Keilin on 14 June 1935 (which primarily served to introduce a visitor) Warburg spoke of the courage necessary for the choice of a research problem.

Dear Keilin,

After two years' work we have now found out the mechanism of action and the active group of the coenzymes of red blood cells. We are very pleased with this, as one can never be sure in such work whether one is going to get stuck. Even scientific research requires courage, if not heroism. As children of our times, we must emphasise this.

Kind regards,
Otto Warburg

In 1967 Warburg expressed his views about the best way of training beginners in research. The opportunity came when a Ph.D. student in Philadelphia, Patrick J. Buckley, was instructed to give a seminar on Warburg and to talk in particular about why Warburg was exceptionally successful. Buckley wrote for help in preparing his talk and received the following reply (in English).

Dear Mr. Buckley,

Thank you for your letter of 8 November. If you wish to become a scientist, you must ask a successful scientist to accept you in his laboratory, even if at the beginning you would only clean his test tubes. If you observe carefully what he does, you will learn how discoveries are made.

It is not necessary that you know already much before you go to a master.

The earlier you go, the greater will be your chance to become a discoverer too.

I am, Yours sincerely,
Otto Warburg

Characteristic of Warburg's views on scientific research are the epigraphs with which he introduced his books. They also give an indication of the range of his reading:

'The most important step in the progress of every scientist is the measurement of quantities. Those who are satisfied with the mere observation of facts have occasionally rendered services, inasmuch as they drew the attention of other people to phenomena which they had seen. The great advances in science are achieved by those who endeavour to find out how much of each there is present.' (James Clark Maxwell, *Theory of heat*, pp. 74–5. (1877) [504]).

De fixé les objets longtemps sans être fatigués.[508]

Goethe to Eckermann: 'How much effort and how much expenditure of mental energy is necessary in order to see a great wholeness as an orderly unit and how important is a quiet, undisturbed time in one's life.'[507, p. 7]

I describe in what follows, exactly and with all the detours, the circuitous route by which this discovery was made so that the criticism which Maxwell (53) made about Ampère, is not levelled against me: That when he had built up a perfect demonstration he removed all traces of the scaffolding by which he had raised it.

Scientific writing

Warburg's scientific prose was highly characteristic of the man. It was distinguished by an exceptional clarity achieved by simplicity, brevity, and rigid scientific logic. Occasional striking and colourful passages enlivened the text. Thus, when he discussed the measurement of the 'action spectrum' of the carbon monoxide compound of the oxygen-transferring enzyme within living cells he wrote[505],

> If one uses reagents which react only with the enzyme and not with any other cell constituents then these cell constituents do not interfere any more than the wall of a test tube interferes with a chemical reaction. Hence enzymes can be studied like pure substances and from their reactions conclusions can be drawn about their composition.

His aim was to formulate conclusions reached from his experiments with as few qualifications as possible and he pursued his experiments until this was achieved. Too many qualifications, he felt, detract from the value of any statement. In his eagerness to be clear and straightforward he occasionally tended to over-simplify. This aroused some criticism, but the simplification was somewhat deliberate, for he subscribed to the view of Bayliss (6) that,

> Truth is more likely to come out of error if it is clear and definite, than out of confusion, and my experience teaches me that it is better to hold a well understood and intelligible opinion, even if it should turn out to be wrong, than to be content with a muddle-headed mixture of conflicting views, sometimes called impartiality, and often no better than no opinion at all.

Scientific controversies

In his day-to-day contacts in the laboratory, Warburg was always friendly and helpful, as long as he was treated with due respect, but this came naturally to his collaborators. He was open, and freely discussed science, public affairs, literature, and people. He was outstanding in his generosity to his collaborators. Many papers to which he had made the main research contribution and had in fact written, did not carry his name, except perhaps in an acknowledgement by the author. A review of his work which established the oxygen-transferring catalyst of cell respiration as an iron porphyrin ended, 'In concluding I wish to emphasize that the results which I have presented are largely due to the work of my collaborators, Drs Negelein and Krebs' [144], yet the whole work was conceived by Warburg himself and he had carried out the greater part of the critical experiments with his own hands.

But when it came to scientific argument Warburg was a very outspoken and dogged fighter for what he regarded as the truth. It was his aim to live up to the standards set by his teachers, Emil Fischer, Nernst, his father, and van't Hoff—to be completely objective, personally detached, and scrupulously truthful. 'Sachlichkeit und Wahrheit' were his ideals. His teachers, he said, would throw out any collaborators who did not come up to the proper standards of reliability and good behaviour. In his controversies he crossed swords with many of the leading authorities of his time, Willstätter, Wieland, Keilin, von Euler, and Weinhouse among them, intensely enough to provoke comment in *The Lancet*.(43) In these disputes he was often scathing, sarcastic, and deliberately ridiculing. He believed that it was dangerous not to contradict erroneous criticisms. He quoted in this context McMunn who in 1885 discovered in animal tissues what is now called cytochrome.[506, p. 61] McMunn was severely criticized by Hoppe-Seyler (a leading authority in the field) who argued that haemoglobin is the only red substance in animal tissues and that the 'histohaematin' of McMunn was an artefact derived from haemoglobin. McMunn did not take up the argument and, as a result, progress was delayed for over 30 years, until Keilin, independently, rediscovered the cytochrome.[21]

To give examples of the way Warburg attacked his opponents: on the question of whether peroxidase (now known to be an iron porphyrin) contains iron, Warburg had stated that some observations, especially the inhibition of the enzyme by cyanide, suggested the presence of iron. This was disputed by Willstätter who found no strict parallelism between enzyme activity and iron content during the early stages of purification. Warburg commented, 'With equal justification Willstätter might have concluded from the C- or N-content of his preparations (which also failed to go parallel with the activity) that the enzyme contained no carbon or nitrogen.'[506, p. 54]

In an obituary of the distinguished physicist, James Franck, Kroebel (42), wrote, 'Unfortunately, Franck's work on photosynthesis was clouded by a quarrel with Otto Warburg which took an unpleasant form. It concerned Warburg's measurements of the quantum yields which, according to Franck's theoretical physical considerations, could not be correct, so that it still remains to be clarified what precisely was measured in Warburg's experiments.' Warburg's reply contained the following passages [466]:

> What 'was measured' was the light energy absorbed by the green algae and the oxygen released because these two values are necessary for the calculation of the quantum yield. By contrast nothing 'was measured' by Franck, neither the light energy absorbed nor the oxygen released. Franck merely calculated the quantum requirement of photosynthesis on the basis of his theories. Thus Franck attempted to guess how many quanta are necessary to release one molecule of oxygen. His result was that eight quanta are required, and when our measurements showed that only one quantum is needed the theories of Franck had to be abandoned.
>
> Kroebel was correct in writing that our measurements clouded the work of Franck but he is not right in calling our measurements 'unpleasant quarrel'. What could be more pleasant in science than experiments which dispose of false theories?

However, the great majority of experts in the field did not accept this reply as a final comment and agreed with Kroebel that what Warburg actually measured required classification.

In 1956 Sidney Weinhouse (71) discussed the significance of the high aerobic and anaerobic formation of lactic acid by cancer cells and the failure of respiration to prevent aerobic glycolysis. Warburg's reply [379] contains the following comments:

> Weinhouse is not appreciative of many important discoveries made since 1925 . . . Most of the comments by Weinhouse might have been written before 1925 . . . Obviously, nothing could be less enlightened than the opinion of Weinhouse that the respiration of cancer cells is high, or even higher, than the respiration of normal growing cells.
>
> Weinhouse dislikes the statement that the shifting of energy

production from the aerobic to the anaerobic state is the cause of cancer. He feels this far too simple: how can cancer, as mysterious as life itself, be explained by such a simple physicochemical principle?

Yet this feeling is not justified. The problem of cancer is not to explain life, but to discover the difference between cancer cells and normal growing cells. Fortunately this can be done without knowing what life really is. Imagine two engines, the one being driven by complete and the other by incomplete combustion of coal. A man who knows nothing at all about engines, their structure and their purpose, may discover the differences. He may, for example, smell it.

In fairness I should emphasize that Weinhouse's assessment is accepted by most experts.

Warburg paid little attention to the suggestion that this kind of controversy was unfruitful. Rationally he was fully aware of the futility of emotional controversy. More than once he quoted Max Planck's dictum (58), 'A new scientific truth does not triumph by convincing its opponents and making them see the light, but rather because its opponents eventually die and a new generation grows up that is familiar with it.' And for the epigraph of his book *Heavy metals* Warburg [506] chose to quote from Pasteur's reference to Berthelot; 'Comment n'a-t-il pas senti que le temps est le juge souverain? Comment n'a-t-il pas reconnu que du verdict de temps je n'ai pas a me plaindre?' This referred to a controversy over the nature of alcoholic fermentation—whether living organisms are necessarily involved as Pasteur had claimed, or whether fermentation is brought about by soluble enzymes as argued by Berthelot and some others.

Warburg's polemical skirmishes did not spring from doubts about the validity and reproducibility of his experimental findings. They arose over the interpretation of his results, in particular of their full significance. He provoked such criticism because he tended to claim that his were the definitive solutions to the problems in hand, when in fact they may have been partial solutions only.

Also characteristic of his feelings towards fellow scientists was his reaction when he had the unusual experience of reading his own obituary in *The Times* of 12 January 1938. An error had occurred; he had been mistaken for a distant relation, the distinguished botanist and Zionist leader, also called Otto Warburg (1859–1938). The error was corrected on the following day with appropriate apologies. Warburg was a subscriber to *The Times* and was both mildly amused and mildy annoyed to read the news of his demise; annoyed because he thought that what he considered to be one of his most important discoveries had been deliberately omitted (that of nicotinamide in the co-enzymes of dehydrogenases and the chemical nature of its reversible reduction and oxidation). He commented, in a letter to me, 'It looks as if the obituary was written jointly by poor old Heinrich Wieland and poor old Torsten Thunberg.' These two scientists had published much work on dehydrogenations, without establishing the chemical mechanism and, quite wrongly, Warburg looked upon them as his adversaries.

One doubts whether he would have derived any more satisfaction from the scanty and largely irrelevant obituary in *The Times* of 4 August 1970.

Warburg's litigiousness and concern about being recognized as right also expressed itself sometimes in quite gratuitous and pompous digressions in his experimental papers, such as, 'We do not hesitate to express our satisfaction that truth once again has won a war. This time it was a 46 years' war on bioenergetics.'[497] His outlook on controversies in science is well expressed by remarks he made in the recorded interview of 1966: 'Since every real discovery in science means a revolution where there are victors and vanquished each of our discoveries has evoked a long and bitter fight. All were eventually decided in our favour.'

Somehow he could not resist asserting himself in this way, although it was quite unnecessary. How different he was in this respect from other scientists. Compare him with Einstein, whose attention was drawn in 1953 by Max Born to comments of Edmund Whittaker who, in the view of most scientists, wrongly attributed the discovery of relativity to Lorentz and Poincaré. Einstein replied, 'Do not worry. Everybody must behave according to what he thinks right. If he convinces others, that is their business. I found my own efforts very gratifying but I do not consider it a sound proposition to defend my few results as "property". I am not at all angry with him. After all, I need not read the stuff.'(17)

Warburg, on the other hand, was quick to anger and would accuse his fellow scientists of being prejudiced, even dishonest—dishonest because he believed they stood by out-of-date ideas because they felt it dangerous to disagree with people in positions of power. Warburg considered that he suffered the same kind of difficulties in winning recognition as had Galileo, of whom he had read in Andrade's contribution to *Notes and Records* (1):

Besides his wonderful genius for scientific discovery by observation and experiment, which led to the replacement of the authority of the ancients by the methods of modern science, he had a notable gift which was the source of all his troubles: he was a master in the art of polemic and welcomed controversy, which Newton so conspicuously shunned. It was his mordant and aggressive argument that led to his misfortunes. It is not in general dangerous to be apologetically, or just quietly right, but to be right and to insist caustically and convincingly that you are right in the face of firmly established authority may be fatal, as Galileo found.

Warburg's own assessment of his work

In 1961, Warburg summarized his achievements and the reasons for his great success in the introduction to a volume of his collected papers.[508] It is quoted here in full because it is a fair summing up, and because it reveals other facets of his personality. It conveys his clear and forceful style of writing; it illustrates his tendency to oversimplify complex issues and to make somewhat exaggerated claims, e.g. 'the de-differentiated growth of cancer cells was explained by a type of energy transformation at a time when the atmosphere of the earth did not yet contain oxygen—a result which explains the ultimate cause of cancer by the anaerobic past of life.' It is also evident that the summary was written by a man who did not hide his light under a bushel.

As a co-worker of Emil Fischer, I prepared the first optically active peptides in the years from 1903 to 1906 and, as one who belonged to the inner circle of his co-workers, I came to know the methods and experimental techniques of this great organic chemist of our times. Later on I began work on the quantum requirement of photosynthesis in the radiation laboratory of the Physikalisch-Technische Reichsanstalt (National Physical Laboratory), where my father Emil Warburg was president from 1906 to 1922. From the experiments in this laboratory, Max Planck derived his law of radiation and calculated 'the quantum of action', $h = 6.55 . 10^{-27}$ erg.sec. In the same laboratory Emil Warburg measured the first quantum yields of photochemical reactions.

Thus it came about that in the Kaiser-Wilhelm-Institut für Zellphysiologie the methods of two widely separated fields, those of radiation physics and organic chemistry, were united from the 1920s onwards in my one hand. The result was the elucidation of the chemical mechanism of enzyme action—the very fast intermediate reactions of protein-bound active groups that we were able to separate from the enzymes and to crystallize. Coenzymes and many vitamins proved to be nothing other than free active groups of enzymes.

Here are some examples to illustrate the interplay of physical and chemical methods characteristic of our institute.

By the methods of radiation physics the substance was discovered that reacts in the living world with molecular oxygen. By the methods of organic chemistry this substance (which is an iron compound) was isolated and crystallised. By the methods of radiation physics the active groups of the hydrogen-transferring enzymes were discovered, and by the methods of organic chemistry they were separated and crystallised. The study of the interaction of the oxygen- and hydrogen-transferring enzymes gave us the respiratory chain—the general solution of the problem of Lavoisier's respiration. By the methods of radiation physics the main reaction of fermentations was recognized as a hydrogen transfer by nicotinamide, by the methods of organic chemistry 1,3-diphosphoglyceric acid has been isolated from yeast, the reactions of which explain the production of phosphate energy in fermentations. By methods of physics the fermentation of cancer cells has been discovered and the de-differentiated growth of cancer cells explained through a type of energy transformation which prevailed at a time before the atmosphere of the earth contained oxygen—a result which explained the ultimate cause of cancer by the anaerobic past of life.

The methods of radiation physics were employed to measure the quantum requirements of photosynthesis. It was found that in *Chlorella* somewhat less than three light quanta are necessary to split $1CO_2$ into $C+O_2$; this means that in red light there is an almost complete conversion of light energy into chemical energy. In this energy transformation 1 light quantum splits off 1 molecule of oxygen from the photolyte which is a carbonic acid transformed by the energy of respiration; $\frac{2}{3}$ of the produced C and O_2 enter into a back reaction so that, in the net balance, three light quanta split one molecule of carbon dioxide. Thus the problem of the multiquanta reactions no longer exists in photosynthesis. The process of photosynthesis is governed by the same principles as the ordinary photochemical reactions of the nonliving world. On this basis, the special chemistry of photosynthesis is nowadays investigated with the methods of manometry and organic chemistry; our findings in general energetics have thereby found excellent confirmation.

The best physical and chemical methods, however, would not have been able to advance cell physiology, had they not been combined with a simplification of the biological technique. Instead of measuring the flow of blood through the surviving organs, the use of thin tissue slices and of single cells was introduced. Red blood cells—nonnucleated from mammals, and nucleated from birds—were the first materials used to investigate the chemical mechanism of cell respiration. Instead of using green leaves to study photosynthesis, unicellular green algae were introduced; the algae were sometimes replaced by the green grana of spinach chloroplasts or the even smaller green grana of *Chlorella* produced by exposure to sound waves. In the investigation of cancer cells, ascites cancer cells were substituted for tumours; whole animal cells were replaced by the grana of liver cells as early as 1914. In all this experimentation it was always our aim to simplify techniques and to attain a speed and precision comparable to the methods of volumetric chemical analysis. With complicated methods we have never discovered anything significant.

To supplement this self-assessment, here are further extracts from the autobiographical tape recording of 1966:

I was born in 1883 in Freiburg in Baden where my father was Professor of Physics. My father came from the Danish branch of the Warburg family and became a Prussian citizen only in 1866 when he was 20 years old. My mother came from a Baden family of Civil Servants and officers; her brother died in action as a general during World War I.

In 1896, when I was 13, I went with my parents to Berlin where my father became Professor of Physics at the University and where later, from 1905 to 1922, he was President of the Physikalische-Technische Reichsanstalt as successor to Hermann von Helmholtz. I grew up in Berlin-Charlottenburg in two official residences, more palaces than houses. Both had been built to the ideas of Frau von Helmholtz. After 1896 I never left Berlin, except during the First World War.

In my parents' home I met the giants of the natural sciences who at that time were drawn to Berlin: the chemist Emil Fischer, the physical chemist Jacobus Henricus van't Hoff and Walther Nernst. From them I learned physics and chemistry: chemistry in the laboratory of Nernst and physics in the radiation laboratory of the Physikalisch-

Technische Reichsanstalt where, under my father's direction, I measured the quantum yield of photochemical reactions.

As a student of physics and chemistry I was already becoming interested in the processes of life, and after I had obtained my Ph.D. degree under Emil Fischer in Berlin, I decided to study medicine in Heidelberg under Ludolf von Krehl. I shall always consider it a great stroke of luck that precisely at the time that my apprenticeship in physics, chemistry, and medicine had been completed, the Kaiser Wilhelm Institute for Biology was founded in Berlin and that Emil Fischer, the Vice-President of the Society at that time, invited me to become a Member of the Institute of Biology. Ever since—that is 53 years—I have stayed at the Kaiser Wilhelm Gesellschaft where I have been able to devote all my time to research, since 1931 in my own Institute, which our visitors from abroad called 'the Palace of Cell Physiology'. I have no doubt that my scientific successes are very largely due to the exceptional measure of freedom and independence which I enjoyed in the Rockefeller Institute of the Kaiser Wilhelm Gesellschaft.

The only interruption to my scientific work during those 53 years were the four years of World War I.

Ever since I began to work independently it has been, and today still is, my aim to find out to what extent the processes in living organisms can be resolved in terms of physics and chemistry. In the course of this I discovered the chemical nature of enzymes, the main tools of life, which Willstätter said, as recently as 1930, could not be explained on a chemical basis. I discovered the chemical mechanism of cell respiration, the chemical mechanism of hydrogen transfer in living cells, and thus the mechanism of all fermentations. I discovered the quantum chemistry of photosynthesis and finally, in the field of medicine, the general and direct causes of cancer.

In support of his fight to obtain recognition for his views, Warburg liked to demonstrate his key experiments and in the early 1950s he set aside a special room for demonstrating his most important methods and discoveries, especially in the field of photosynthesis. His own description which stresses the historical perspective of his work runs as follows [370]:

Recently we set up in our laboratory a room in which some important methods and discoveries in the field of photosynthesis are demonstrated to everybody who is interested.

We demonstrate the use of the special bolometer which was developed by Lummer and Kurlbaum towards the end of the last century at the Physikalisch-Technische Reichsanstalt at

Charlottenburg, the same laboratory where in 1899 Lummer and Pringsheim measured bolometrically the radiation of black bodies and obtained the data from which Max Planck in 1900 calculated the value of the action quantum *h*, where Emil Warburg—between 1915 and 1930—measured for the first time the energetics of photochemical reactions, and where Otto Warburg—in 1920—for the first time measured the energetics of photosynthesis.

We also demonstrate our manometry, which measures the effects of light on photosynthesis. It is the same manometry through which we discovered the oxygen-transferring iron of respiration, the yellow enzymes, nicotinamide and the mechanisms of biological hydrogen transfer, the oxidative reaction of fermentation and thus the chemical mechanism of fermentation—in short, a large part of the biochemistry of today. We demonstrate also the cultivation of a *Chorella* which utilizes light maximally, without which even perfect bolometry and manometry are useless. In this field also, that of *biological* methods, we have the earliest and longest experience: it was *Chlorella* which we introduced as experimental material for photosynthesis research in 1919 and since then all relevant papers in the field have been carried out on *Chorella*.

Anyone who has learned in our laboratory our three essential methods—bolometry, manometry and cultivation of *Chorella*—can himself carry out the most important experiments of photosynthesis; he can see how one quantum of light generates one molecule of oxygen from *Chorella*; how in the dark two-thirds or more of the oxygen evolved in the light back-reacts through respiration; how photosynthesis decreases in monochromatic red light but increases in blue–green light; he can measure the action spectrum of blue–green light and he can see and learn many other things.

At a relatively early stage of his career, Warburg became fully aware of his outstanding ability. He knew that in his field of biological chemistry he towered above the great majority of even the leading scientists of his generation. He felt that he carried on where Pasteur (who, like Warburg, was primarily trained as a chemist and who applied his knowledge to biology) had left off. Long before he received his Nobel Prize he thought that he had earned one—unlike many other laureates who, though conscious of their ability, would not place themselves above many other candidates. When this award was announced in 1931, his reaction was, 'It's high time.'

Once when a colleague mentioned to him that he had taken

part in a discussion on whether Warburg or Otto Meyerhof (his pupil) was the greater scientist, Warburg was taken aback and asked whether this was really a question which required discussion. But a little while later, after reflecting upon it, he commented 'Meyerhof war vielleicht klüger, aber ich konnte mehr'. This is a remarkably detached, reasonable and modest assessment.

Of his greater skill (*Können*) there can be no doubt. Meyerhof's technique in the laboratory was not particularly good, whilst Warburg was a genius in devising and handling techniques. Meyerhof's intellectual interests may have covered a wider area, but I do not think his intelligence was greater. Warburg's skill was not merely technical but highly intellectual. Of course their intelligences were not of the same kind.

Teaching and committee work

Warburg took little interest in teaching or in educating a new generation of scientists. When a colleague once tackled him on the importance of teaching and in particular of transmitting a tradition of attitudes to the scientists of the future, (implying that this would be a worth-while job for Warburg), he received the reply, 'Look, Meyerhof, Theorell, and Krebs were my pupils. Have I not done enough for the next generation?'

Warburg never taught undergraduate courses. His rare lectures were addressed only to fellow scientists. Seldom would he attend scientific conferences and he declined the great majority of lecture invitations because he deprecated the widespread practice of 'academic tourism'. Accepting too many opportunities of travel, he said, is the productive scientist's way of perishing honourably—but perish he will, especially the senior scientist.

He kept aloof from committee work, because he regarded much of it as a waste of time. After the Second World War, his advice was often sought and readily given, especially within the councils of the Max Planck Gesellschaft, but he tended to be cross when his advice was not accepted in full and he restricted the time he gave to such work to a minimum.

Eccentricities

Warburg was in many ways an eccentric—as he was the first to admit. Some of his eccentricities concerned food. Once he was convinced that cancer could arise from a large number of chemicals if there are long periods of exposure, he became extremely careful to avoid food which had been treated with special chemicals ('additives'). He had heard that bakers tend to get eczema and he thought that this had something to do with the bleaches added to the flour. So he would never eat bread from a shop if he could help it. For at least fifteen years or so he would eat only bread that Heiss baked at home from 'unadulterated' flour. Once a week Heiss had to spend one long evening baking, a skill he had learned at a farmhouse. Warburg installed an electric baking oven in his house and eventually also an electric kneading machine. He was also afraid of artificial fertilizers, insecticides, and pesticides, and bought adjacent allotment land, increasing the size of his garden to 4 500 square metres in order to produce food. He employed a full-time gardener who also looked after the hens, ducks, geese, turkeys, and rabbits. Most of the vegetables he needed were grown in the garden, as well as pears, apples, strawberries, raspberries, red currants, and apricots. Warburg would not permit the use of artificial fertilizers, the horses and poultry provided sufficient manure. Instead of using pesticides, he encouraged the nesting of tits. There were ten nest boxes in the garden and Warburg saw to it that they were well-fed during the winter so that they did not desert the garden in the summer; they were very diligent in catching caterpillars and other pests.

Through the School of Agriculture he obtained milk from a special herd. Thirty litres were supplied at a time; these were centrifuged in the laboratory and converted into cream and butter. Occasionally he permitted the purchase of French butter because in France the use of herbicides and pesticides was more strictly controlled than in Germany. Altogether his whims, fancies, and anxieties about food were at times rather exasperating to Heiss, who did all the cooking.

Another eccentricity was his somewhat exaggerated love for everything English. Ever since he first visited Cambridge before the First World War he had been an Anglophile. He loved

England because, he once said, 'the English tolerate headstrong and eccentric (*eigensinnige*) people—like me.' He also loved the keeping up of traditions and the dignified pomp and circumstance of the ceremonial which he experienced when he was given an honorary degree at Oxford in 1965. 'England is the last bastion of the old Europe', he said. Earlier in his life he travelled to England about twice a year for brief periods to buy his suits in Savile Row, to buy riding kit and above all antique furniture. He surrounded himself in his home with beautiful pieces of old English furniture. He held when something was English it could not be wrong, and was a long-standing subscriber to *The Times*.

Weaknesses

Warburg's exceptional qualities of intellect and character were liable to be influenced by his deep emotions which also on occasion tended to cloud his judgement. His intellect enabled him to approach his chosen field of work with sharp, constructive, and critical reasoning and with wide-ranging imagination and resourcefulness. His character qualities enabled him to pursue his aims with dedication and discipline and to avoid the distractions which tempt successful scientists into the 'corridors of power', to travel and to waste time on being lionized.

Such qualities were the fount of his outstanding achievements because emotional factors did not enter his routine day-to-day research. Emotions came to the fore when his work—his conclusions, not his experiments—were criticized by others, or when he thought, often wrongly, that people did not treat himself and his work with due respect. Then he was liable to be touchy and resentful, and he could never forget or forgive even minor quarrels. Many people—including many admirers—suffered. Quite wrongly he would accuse people of lacking intellectual honesty or simply of incompetence, choosing to misinterpret a comment or to imagine a slight. Humility, the greatest of all balancers, was not at all his forte. These weaknesses, combined with an urge to assert himself, led sometimes to a failure to acknowledge other points of view and to fruitless controversy. It was a failure which, alas, brought upon him an intellectual isolation, and so it came about that a distinguished biochemist, Efraim Racker (60), was prompted, not without justification, to write in 1972, 'Few will challenge the statement that Warburg was one of the great biochemists. His experimental approach was monumental and ingenious. Yet Warburg's views on the two vital areas of his research interests, cancer and photosynthesis, are now almost universally dismissed as erroneous and naive.'

Such comments, it must be emphasized, refer to views, not to experiments. Warburg rightly prided himself that all his results were readily reproducible. The execution of his experiments was never influenced by wishful thinking though he could sometimes forget all about other people's experiments and arguments if they did not suit his own theoretical concepts. Another weakness, then, was his passion for his own theories, a passion which on occasion blinded him.

One suspects that some of the eccentricities in his personal life were also based on judgements distorted by emotion. His food fads arose from exaggerated fears of cancer and some of his other whims from an overanxiety to protect and improve his health.

Neither his obsessions about supposed 'enemies' nor his occasional erroneous judgements detract in any measure from the greatness of Warburg or his work. The greatness of a personality rests on positive, outstanding achievements. Every human being has his flaws. As long as these do not interfere with his achievements, they can be ignored in the final assessment of the importance and value of the man. But the biographer must include mention of them if he is to convey a genuine and not an idealized picture of the whole personality. What the picture loses in respect to perfection, it gains in respect to truth and humanity. Otto Warburg was a man who made full use of his gifts in the interests of scientific research. Many a highly gifted person does not make full use of his gifts. With some this is deliberate; with some, perhaps, it is because of lack of strength of character.

Roots of Warburg's personality

Warburg's personality was no doubt powerfully shaped by his family background—through both heredity and environment.

Many of his ancestors were hard-working, self-disciplined, public-spirited, and devoted to learning and other intellectual pursuits. Several of his close relations have been described also as strong individualists, self-contained, and not readily forming warm personal friendships.

A contemporary relation, the art historian Aby Warburg, was characterized in obituaries by passages which would also fit his cousin Otto: 'Unrelenting against compromise and half-measures, a fighter, a judge courageous and severe, a servant of scholarship. So powerful was his personality that he will live on as an inspiring example.' Another writer thought that Leonardo da Vinci's saying, 'He never turns back who has found his star'[22] was a fitting motto for Aby Warburg's life. It also fits Otto.

His father's style of scientific writing was a model for Otto, aiming at maximum brevity and clarity as exemplified by Emil Warburg's classical textbook. Otto's distant cousin Siegmund (70) coined for his personal bookplate the motto, 'Progress in thinking is progress towards simplicity'—which also reflects Otto's style, as mentioned on page 63.

Emil Warburg was also Otto's model for high standards of conduct and dedication, as may be gathered from what James Franck (22), one of Emil's distinguished pupils, said in an obituary address, (the same James Franck whom Otto Warburg later attacked in an acrimonious controversy):

> We admire him as the master of the art of experimentation whose results greatly enrich our science; as the teacher, whose pupils continue his work in his spirit all over the world, in academic centres as well as in industry; and last but not least as the man whose objectivity, fairness and lack of prejudice was never affected by success whether his own or that of others.
>
> His classical achievements did not spring from the sudden flashes of an exceptionally gifted brain. They came out of the hard work of a straightforward and objective experimenter who knew the art of asking questions ripe to be tackled and who thereby arrived at so many valuable and clear-cut answers.
>
> In 1895 he succeeded Kundt in the Chair of Physics at Berlin, but before this he had already published the first edition of his *Experimental Physics* which presented concisely and precisely the substance of his great course of lectures for beginners. From this book,

many generations of students, students of physics in particular, have learned the foundations of physics and from it have acquired a high standard of simplicity and brevity in expression together with elegance and precision in definition. The book is not meant for 'dipping into'; if one wants to absorb its lessons one has to study it in depth because every word of every sentence has been carefully thought out. The book is characteristic of Warburg's own attitude to science which he also expected, almost as a matter of course, from his pupils. He who is not committed heart and soul to research had best leave it alone. His successes have justified his views, for his book has so far appeared in 22 editions. The last, which appeared in 1929, was brought up to date by Warburg himself, who at that time was 84 years old.

He who was not full of the desire to learn and was not prepared to devote all his energies to the work did not attempt to ask Warburg to be admitted to his laboratory. Once accepted, you entered a community almost like a big family, for life was lived almost entirely in the laboratory. You appeared punctually in the morning and certainly well before the professor made his round. At mid-day there was only time for a brief snack in a nearby bar, and for the evening meal people preferred to stay in the laboratory where work often continued late into the night.

The long working day made it necessary to have occasional breaks, during which there were discussions about what the professor had said and when individual problems were aired and thrashed out. These discussions were no less important than the bench work. There were no secrets in Warburg's laboratory any more than there was envy when others made good progress. The objectivity of the leader transmitted itself to every member.

Incidentally Emil Warburg's family did not always agree with his single-minded dedication to his professional work. His daughter Lotte once made the following remark about her father in her diary: 'Papa is not interested in people. When one tries to tell him something about someone, he says, "What has that got to do with me?" ' When talking about their very large official residence, she and her two sisters commented, only half-jokingly, 'Papa doesn't even know where Mama's bedroom is.'

Otto had many of his father's traits. Charming and helpful to those who shared his interests, he was very cool, indifferent, and even unfriendly towards those who did not interest him or would 'waste his time'. So it came about, as is not infrequent with strong personalities, that he struck different people in very different ways, depending on whether or not they managed to strike a chord with him. In his determination to concentrate his efforts on his chosen activities, he showed little interest in and warmth towards people not connected with those activities.

Warburg referred on several occasions to the debt of gratitude he owed his teachers. When, in 1963, he was asked to write an autobiographical chapter for *Annual Reviews of Biochemistry*, he wrote in his introductory paragraphs:

> The most important event in the career of a young scientist is personal contact with the great scientists of his time. Such an event happened in my life when Emil Fischer accepted me in 1903 as a co-worker in protein chemistry, which at that time was at the height of its development. During the following three years I met Emil Fischer almost daily and prepared under his guidance the first optically active peptides.

What he owed to Fischer and other teachers was not a direct introduction into his specific area of research and methodology but a style of researching and an attitude towards science as a profession. Customarily a teacher helps a beginner to choose a subject and to acquire technical skills. In this sense Warburg cannot be said to have had a teacher because he himself formulated the problems on which he wanted to work, and developed independently the methods he required.

Under the influence of the parental home and the scientists he met there, Warburg arrived at the view that too many university professors were not practising science for science' sake but rather for reasons of prestige or financial reward—the latter particularly in medicine where the title 'Professor' opened the door to a lucrative private practice. He saw that many university teachers abandoned research as soon as they reached the upper echelons of the profession. For them, he said, science was merely a means to an end. He had learned from Emil Fischer, from his father, from Nernst, from Einstein, and from other members of the Berlin circle, to condemn this attitude. Warburg devoted himself to science for science' sake. In accordance with the traditional views (which go back to Francis Bacon) he was convinced that to extend knowledge is worthwhile and contributes to the well-being of humanity.

Warburg believed, with much justification, that in some branches of German academic life, personal connections and favouritism played a much greater role than real ability and quality of character. This he considered as unimportant in mathematics and physics, of relatively little importance in chemistry, but of major importance in medicine and some of the humanities. At the interview which led to my appointment, he said, 'I cannot help you towards an academic career. I can teach you something but I have no connection with the University of Berlin. I am an outsider and they don't like me at the University. If you want to make your career in a university, you had better cling to the coattails of some old ass of a professor.'

He referred to this kind of 'corruption' very frequently during our informal laboratory conversations, and on one occasion he illustrated it with an account of how he himself had become a 'Professor'. As a Member of the Kaiser Wilhelm Gesellschaft, with no connection with the university, he had no right to this title. However, the Minister of Education, acting on the suggestion of influential academics, would occasionally award the title to people of special merit who did not belong to a university. In Warburg's case, it happened that while serving in Russia during the First World War it was in his power to requisition animals for food, and as a token of gratitude he sent Emil Fischer half a sheep because food was scarce in Germany. The only reply he received from Fischer, which came a few weeks later, was a telegram congratulating Warburg on being awarded the title of Professor.

Warburg emphasized that this incident in no way cast doubts on Fischer's integrity. Warburg told the story as a criticism of the system which made it possible for an influential person, through a letter or telephone call to an official in the Ministry of Education, to obtain advancement for one of his pupils. Warburg believed that this power was often misused, but there was in his own mind no doubt that there had been no misuse in this particular case.

This theme of 'corruption' appears to have been a topic of conversation in the parental home, for there is extant some satirical doggerel by Otto's sister Lotte. Lotte was skilled with the pen and occasionally contributed to the *feuilleton* of the

Neue Zürcher Zeitung. I find these verses difficult to translate; (they are reproduced in the German version of this biography, p. 102). They deal with the way in which careers were not infrequently made in German universities. This is their message:

> The aspiring academic need not be a competent scholar: professorships can be achieved another way. The aspiring young man has only to ingratiate himself with his professor—and his wife—by courting them with flattery and servility. He must cultivate influential people and if not already related to the professor by a family relationship, he must show a keen interest in the professor's daughter. Publications are, of course, necessary, but need not be original: rehashing other workers' papers will be quite sufficient. But the really important thing is to win the favour of people in influential positions.

This critical assessement did not refer to mathematicians and physicists and only marginally to chemists. Otto often expressed the view that people who had reached the top by these 'back stairs' played a major role in the humanities and in medicine.

This kind of favouritism which Warburg and his family censured occurs of course to some extent in every society. What they criticized was excessive, wide, and overt favouritism. In Britain, the great majority of appointments are made by committees of a sufficient number of people to see to it that justice is done. I have been a member of such committees in well over a hundred cases and I believe in every case a genuine effort was made to appoint the best person. Errors of judgement have occurred but that is a different matter. Nepotism probably plays a rather smaller part in Britain than does the 'old school tie'.

Jakob Heiss

A special acknowledgement must be made to Warburg's longstanding companion, Jakob Heiss, who after his demobilization in 1919, came to keep house for Warburg on the recommendation of his military friends. Heiss was a bachelor. He was born in 1899 and during the First World War served in a Prussian infantry regiment of the Guards. Out of the initial servant–master relationship grew a close personal friendship. Warburg and Heiss became virtually inseparable, not only during the daily routine but also during leisure hours and holidays. Heiss later became the unofficial secretary and manager of the Institute—unofficial in that until very late in life he was not paid by the Institute, although he spent much time in its service. Heiss was an understanding, devoted, and self-denying friend who did everything he could to support and protect Warburg—not always an easy companion—and to make his life as congenial as possible. He also did much in calming and appeasing Warburg's emotions when they were raised by controversy and resentment. For though caring for Warburg most considerately, Heiss would not hesitate to express his own independent and often shrewd views. Warburg valued his common sense and would often defer to Heiss's opinions.

The fascination of Warburg's personality

Warburg made a deep impression on all who came into contact with him, irrespective of whether or not they approved of his views. Whenever people who knew him met, conversation sooner or later turned to the subject of his personality. Apart from his penetrating intelligence, what fascinated people were his intellectual honesty and straightforwardness, his singularity of purpose and his industry, the wit and humour of his pertinent comments on affairs (scientific and other), his generosity in helping people in the laboratory, and his oddities and extravagances. Those who knew him well were very ready to overlook his weaknesses, his touchiness, his resentfulness, his prejudices, and his harshness against those whom he unjustifiably regarded as his 'enemies'. Thus David Keilin, despite having been rather abused and ridiculed by Warburg, was keen to support his nomination to the Foreign Membership of the Royal Society.

Of Warburg the scientist it may be said, as Cuvier (14) said of Humphry Davy (when comparing him with other leading scientists): 'He rose like an eagle and from high above illuminated a large area of science with a bright torch. Others modestly shone their little lights on limited objectives. His name stands at the head of many chapters; the names of others appear in single paragraphs.'

Of Warburg the man it may be said, as Goethe said of Faust, he was one of those 'who never ceased to strive.' And thus, like Faust, he was a good man.(25)

Notes

1. Max Warburg, one year younger than Aby, wrote in his autobiographical notes (private edition 1952, copyright of Eric. M. Warburg, pp. 5–6) about the arrangements with his brothers:

> Aby, as the firstborn, was expected to enter the banking business and eventually become a partner, as was the tradition of the firm. However, he refused to become a banker and instead studied the history of art at the Universities of Strasbourg and Bonn.
>
> When I was 12 years old Aby proposed that I should buy from him his birthright, not in exchange for a 'mess of pottage' but in return for my undertaking that I would pay for the purchase of all his books. I was a child, and the bargain seemed excellent to me. My father's business, I thought, would surely be prosperous enough to allow me to buy some Schiller, Goethe, and perhaps also Klopstock. We sealed the pact solemly with a handshake. This pact was perhaps the most lighthearted of my life but I have certainly never regretted it because it was the foundation of Aby's library and of his highly successful career. Moreover, his activities have enriched the lives of the whole family, have widened the horizons of each one in a way which none of us could have foreseen.
>
> Aby never ceased working with almost frantic zeal for his library and for his new approaches to the history of art. He did not write very much; sometimes I urged him to write more but he always replied, 'If more books were read, fewer would be written.' He had still many plans when he told me, a few months before his death, 'One must organise one's life as though one would live forever—but one must be ready to die every day.'

This forward-looking, optimistic posture echoes a saying often quoted in Germany: 'If I knew that the world was to come to an end tomorrow, I would still plant a young apple tree today' (attributed to Martin Luther).

2. Käthe Pleuss (1882–1948), Lotte Meyer Viol (1884–1948), Gertrud von Wartenberg (1886–1971).
3. A glimpse into Warburg's childhood is given by an Open Letter of congratulation addressed to him on his 80th birthday by his former playmate and subsequent fellow-biochemist, Karl Thomas. The letter appeared in *Naturwissenschaften* **50**, 629 (1963). Thomas was for many years Professor of Physiological Chemistry at Leipzig.

Dear Otto Warburg,

The editor of *Naturwissenschaften* has asked me to commemorate your 80th birthday with a few personal words. I gladly accede to his request for there is nobody else alive today who has known you so long as I.

Our fathers both came to Freiburg as professors in 1876 and they both married in 1880. Our mothers met once or twice every month on their regular rounds of visiting in their circle of women friends, where they proudly showed off their babies. I am sure, however, that at that time I took no notice of you. That happened a few years later when the Thomas boys played with you and your three sisters, either in your home or in ours. And then, a few years later, we were classmates at a private school which taught us as much in three years as would have taken 4½ at the state school. Then we were in the same form at the strictly Classics-orientated Berthold Gymnasium until we were 12. Other classmates of ours were Fritz Hartung, later historian at Berlin; Otto Koellreuther who became Professor of International Law at Munich; Erwin Poensgen, for many years ambassador to Venezuela; Richard Siebeck, Professor of Medicine at Berlin and Heidelberg—not at all a bad year, I think. We met again as students at Freiburg. You were reading chemistry, I medicine. We went to the same lectures in chemistry and physics, and together we listened to August Weismann lecturing on the origin of species. At that time you were already attracted to the physical-chemical approach, which struck me as being exceptional.

A few years later we met again in the Department of Chemistry at Berlin. You were working on the stereochemistry of leucine with Emil Fischer (who at that time was exploring the chemistry of proteins) while in Rubner's laboratory nearby I was beginning to study the minimum protein requirements of man. After obtaining your Ph.D. degree you went to Heidelberg, to Krehl, to work on the respiration of red blood cells. Often you spent some time at the Zoological Station, Naples, where you discovered the sudden rise in the oxygen consumption of the sea-urchin egg after fertilisation. Later, in 1913–14, we both started work in the Kaiser Wilhelm Gesellschaft, I at the Institute of the Physiology of Work, and you at Dahlem. Then came the war and we both went to the Front. I was wounded and returned to Berlin in 1916. You, I believe, returned only at the end of the war. We were both in Berlin during the revolution and experienced its consequences. Then we became separated for a longer period: I became Professor of Physiological Chemistry at Leipzig while you remained at Dahlem, first as section head of the Institute of Biology, later as Director of your own Institute for Cell Physiology. There you only had a few, but very distinguished, academic collaborators and pupils. I mention Meyerhof, Krebs (today Sir Hans at Oxford) and many others, the last being your pupil Bücher. You were not hindered by having to give lectures and practical classes, as was I at Leipzig, and you could devote the whole day to your research work. We all know that you still continue to do this today. You make each of your technical collaborators practise every technique under your supervision until he reaches the stage where he can improve no further. Then and only then do you allow them to experiment on their own—but only under your detailed instructions and with you standing over them for as long as you feel necessary. No experiment goes on in your laboratory of which you do not know all the details. But you allow your collaborators to publish their work under their own names only.

You told me once that for your life's work you had had the best teachers of the times: for chemistry Emil Fischer with his great laboratory, for physics your own father, and for biology, Krehl with his vitality, his superior intellect and ready communicativeness, at his widely diversified Department of Medicine. But it is well known that the best teachers alone are not of much use unless their pupil pulls with them, knowing exactly what he is aiming at and how his teachers, their examples and advice, can help him on. You have, thanks to your fine scientific education and your own gifts, carried out experiment after experiment, measurement after measurement, and only when all the data were reproducible and in agreement did you publish the result. Every field you researched was a completely new territory for which you yourself created the tools to explore.

Neither of us chose the easy path; we were both fortunate in being able to choose the finest profession we could imagine. It filled our lives completely. We followed our own desires, which differed only in detail, not in substance.

'Nach dem Gesetz, wonach du angetreten
So musst du sein, dir kannst du nicht entfliehen
So sagten schon Sibyllen, so Propheten;
Und keine Zeit und keine Macht zerstückelt
Geprägte Form, die lebend sich entwickelt.'

With cordial greetings for your birthday
Your old friend,
Karl Thomas

The closing verse, well-known in Germany, comes from Goethe's *Orphische Urworte*; the lines are headed 'Dämon'. Goethe's poetry is difficult to translate (see L. Forster, *Penguin Book of German Verses* (1957)). The following gives the sense and scansion but neglects the rhyme.

The Law which governed your beginning
Will always govern you. You can't escape yourself.
So spake the Sibyls long ago—and Prophets too.
Nor time nor force can smash the casting mould
A mould that, as it lives, develops.

4. *The Kaiser Wilhelm Gesellschaft.* The Kaiser Wilhelm Gesellschaft was founded in 1910 by private initiative with the intention to provide ideal research facilities for promising scientists. Behind this was the conviction that scientific research can be of great material and spiritual benefit to humanity.

The carefully selected few were given well-equipped laboratories, freedom from routine teaching duties, and a minimum of administrative chores. The funding did not come from the Government; it came from individuals and from industry. A special inducement to individuals to contribute was an invitation to a dinner given by the Kaiser at the Imperial Palace—for those who gave at least 10 000 marks (then equal to 500 gold sovereigns).

The setting up of the The Kaiser Wilhelm Gesellschaft was the beginning in Germany of paying scientists to carry out full-time research. Up till then, German university scientists, like those in other countries, were primarily paid as teachers, not as researchers. Their teaching duties, however, left some time for research, which they were expected to carry out.

Warburg's section was in the The Kaiser Wilhelm Institute for Biology. The sections of the Institute did not specify an area of biology but were named after their heads—'Abteilung Warburg'. This signified the complete freedom in the choice of research topics.

5. The mansion house at Gross-Kreutz was built in 1763, designed

by leading architects of the time. The last owner was Bodo von der Marwitz. At present the building houses the administration offices of the state-owned animal breeding and seed production centre.

Warburg liked to refer to the house as a '*Schloss*' (stately home), built by one of the generals of Frederick the Great. However, although there were several generals of the name von der Marwitz during the reign of Frederick the Great and later, none had anything to do with Gross-Kreutz. This was one of several instances where Warburg's imagination, fired by snobbery, ran away with him.

6. The Kaiser Wilhem Institut für Zellphysiologie, which had been opened in April 1931, was renamed the Max Planck Institut für Zellphysiologie in 1953. In March 1972 the Max Planck Gesellschaft decided not to continue this Institut because efforts to find a suitable successor had been unsuccessful. The staff and laboratory equipment were placed in other Instituts of the Max Planck Gesellschaft and the building was refurbished to accommodate the library and archives of the history of the Max Planck Gesellschaft. It is now named the 'Otto Warburg Haus'. It was officially opened on 8 March 1978.
7. The visit to Barcroft's laboratory is mentioned by R. Siebeck in *Handbuch der biochemischen Arbeitsmethoden* (ed. E. Abderhalden), Vol. 8, p. 33 (1915). Warburg has briefly described the history of manometry in reference[504].
8. Modern spectrophotometry and its application to biological problems has several sources. It started in Warburg's laboratory in 1928–29 in the course of the work on the absorption spectrum of the oxygen-transferring enzyme (*Biochem. Z.* **214**, 64 (1929)).

Warburg's collaborator, Haas, in 1935 described several improvements to the original apparatus (*Biochem. Z.* **282**, 224 (1935)): he introduced a particularly efficient hydrogen lamp as the light source; with the help of two monochromators arranged in series in isolated narrow bands from the continuous spectrum; and instead of quartz he used fluorspar for making prisms and lenses. At the same time, Ardenne and Haas (*Z. phys. Chem.* **A174**, 115 (1935) improved the sensitivity of the instrument and thus made it possible to measure optically the reduction of cytochrome in a 50 per cent yeast suspension.

Independently, Hogness in Chicago developed a spectrophotometer (*J. phys. Chem.* **38**, 108 (1934); **41**, 379 (1937)). Hogness visited Warburg's laboratory in 1937 and soon afterwards, Haas joined Hogness in Chicago. In 1941 (*J. opt. Soc. Am.*, **31**, 682) Cary and Beckman described their spectrophotometer, which under the name of the 'Beckman spectrophotometer' was commercially extraordinarily successful.

A further decisive advance was the 'double-beam' instrument of Chance (*Rev. scient. Instrum.* **22,** 619 (1951); *Nature, Lond.* **169,** 215 (1952)).

9. Sir Richard Doll estimates the percentage of avoidable cancers at 80–90 per cent.
10. The state of knowledge in the field of biological oxidation at the time when Warburg began to enter the field is described by Marcel Florkin in his essay, 'Early theories of the biological oxidations of intracellular respiration' (*Acta historic. Leopold.* **9,** 59–68 (1975)). The comparison with present knowledge makes it impressively clear how much Warburg pioneered the advances, especially as he demonstrated how speculations and experiments on models could be replaced by direct chemical and physical measurements on biological material.
11. Later concepts on photosynthesis have been discussed by H. T. Witt (*Q. Rev. Biophys.* **4,** 365–477 (1971)). The current state has been reviewed by D. I. Arnon (Photosynthesis 1950–1975; changing concepts and perspectives. *Encyclopaedia of plant physiology*, New Series, Vol. 5: Photosynthesis I. (ed. A. Trebst and M. Avron), pp. 7–56. Springer-Verlag, Berlin (1977).
12. Adolf Butenandt has pointed out (*Naturwissenschaften* **7,** (in press)) that isonicotinamide hydrazide was developed simultaneously and independently by the Hoffmann–La Roche Company at Basle and by the Bayer-Werke in Elberfeld.

    A detailed history of the discovery of the therapeutic action of isonicotinamide hydrazide has been published by E. Krüger-Thiemer (*Jahresbericht des Tuberkulose Inst. Borstel* **3,** 192–4 (1956)). The two companies have published a joint declaration recognizing the simultaneous independent discovery (*Experientia* **8,** 304 (1952)). We are indebted to Professor Butenandt for this information.
13. The following list of honours shows that Warburg's achievements were properly appreciated in many circles. He was the first academic scholar to receive the Freedom of the City of Berlin.

1925 Adolf von Baeyer Memorial Medal of the Verein Deutscher Chemiker
1931 Dr Joseph Schneider Gold Medal of the University of Würzburg
1931 Nobel Prize for Physiology and Medicine
1934 Foreign Membership of the Royal Society of London
1951 Order Pour le mérite für Wissenschaften und Künste

1953 Great Cross of Merit and Star of the Order of Merit of the German Federal Republic (on 70th birthday)
1954 Honorary degree, Technical University of Berlin
1957 Honorary Member of the Gesellschaft Deutscher Chemiker
1958 Great Cross of Merit and Sash (on 75th birthday)
Ernst Reuter Silver Medal, Berlin
Honorary Degree of Medicine, University of Heidelberg
1962 Ludwig Darmstaedter and Paul Ehrlich Prize, Frankfurt
1963 Freeman of the City of Berlin (on 80th birthday)
Harnack Medal (highest scientific distinction of the Max Planck Gesellschaft)
Institution of the Otto Warburg Medal by the Gesellschaft für Physiologische Chemie
1965 Honorary Degree, University of Oxford

Many other honours were offered, which Warburg did not accept. He felt he could not spare the time to receive them in person—a normal requirement in the conferment of honorary degrees.

14. The words 'malum immedicabile cancrum' are taken from Ovid's *Metamorphoses* Book 2, 825.
15. Special skill is required to do justice in classical Latin to modern scientific and technical terms.
16. Amaryllis was a shepherdess. The quotation is from Virgil's eighth pastoral poem. The words are spoken by Alphesiboeus, disguised as a sorceress.
17. After his return to Berlin, Warburg wrote me the following letter:

Berlin-Dahlem
28 June 1965

Dear Sir Hans,

You have certainly felt how delighted I was to see again old England, almost unchanged—the last bastion of old Europe. Obviously you could not have wished for greater luck than to travel to England in 1933; I am convinced of this.

When I drove to the airport I managed to accomplish a good deed by stopping a bolting black stallion of the Household Cavalry, which otherwise would have raced into the cars at Hyde Park Corner.

Kind regards also to Lady Krebs whom, I hope, you will bring to Dublin.

The second paragraph of this letter is puzzling. An enquiry to the commanding officer of the Household Cavalry received the reply that they had no black stallions. Was the incident merely a dream?

The reader might be surprised at the formal nature of the address in this letter. In earlier years Warburg letters to me had begun, as was customary, 'Dear Krebs'. The knighthood title seemed to please him, like everything English.

The reference to Dublin at the end of the letter was in connection with a symposium on yeast metabolism.

18. Simon Flexner (1863–1946) was a distinguished bacteriologist who played a major part in organizing the Rockefeller Institute and later became its Director. He invited Warburg to lecture there in the 1920s and in conversation he was very kind to Warburg, but Warburg thought his manner was unduly patronizing. (Biographical details for 15th ed *Encyclopaedia Britannica* 1975.)
19. See Fraenkel, H. and Manvell, R. (1964): *Hermann Göring*. Verlag für Literatur und Zeitgeschehn GmbH, Hannover, p. 125.
20. Information received in a private letter from A. von Muralt to H. A. Krebs.
21. During the last decades there has occurred an extraordinary change in the style and tone of polemic in science. Until the early part of this century, polemic was often fierce, ill-tempered, sarcastic, and personal, intended in fact to ridicule the opponent. Today polemic is no less outspoken on the technical aspects but is, as a rule, personally gentle and well-mannered. But in some of the humanities the fierce and aggressive tone of polemic has hardly changed.
22. Nò si volta chi a stella è fisso (Richter, J. P.) *The literary works of Leonardo da Vinci*, (3rd edn), Vol. 1, p. 388. Phaidon Press, London. (1970).

People mentioned in the book

This section supplies biographical details of people mentioned in the book. Not included are names of well-known people and of those whose connection with Otto Warburg is made clear in the text.

Andrade, Edward Neville da Costa (1887–1971). British physicist, London.

Ardenne, Manfred von (1907–). German physicist, engineer, and researcher in the field of experimental medicine, Berlin, Dresden.

Arnon, Daniel Israel (1910–). US plant physiologist, Berkeley, California.

Baeyer, Adolf von (1835–1917). Leading German organic chemist, Munich, Nobel Prize 1905, for work on organic dyes and hydro-aromatic-compounds.

Ball, Eric G. (1904–). US biochemist, Harvard Medical School, Boston, Massachusetts.

Ballin, Albert (1857–1918). Director General of the Hamburg Shipping Co. HAPAG (Hamburg–America Paketenboot AG).

Barcroft, Sir Joseph (1872–1947). British physiologist, Cambridge.

Barron, Eleazar S. Guzman (1898–1957). US biochemist, Chicago and Baltimore.

Bayliss, Sir William (1860–1924). British physiologist, London.

Beckman, Arnold O. (1900–). US instrument maker, Fullerton, California.

Behrens, Sir Leonard (1890–1978). Lancashire industrialist. Vice-President Liberal Party Organization. Hon. Member of Royal Manchester College of Music. Vice-President of U. N. Association; Hon. President, World Federation of U. N. Association and Acting President, Stockholm, 1951; New York, 1963.

Bergmeyer, Hans-Ulrich (1920–). Enzymologist and biochemical analyst, built up the industrial production of enzymes for the Boehringer Mannheim Company.

Berthelot, Marcellin (1827–1907). French physician and chemist, Paris.

Born, Max (1882–1970). German physicist, Berlin, Frankfurt, Göttingen, Cambridge, Edinburgh, Bad Pyrmont. Nobel Prize 1954, for work on quantum mechanics.

Boveri, Margret (1900–75). Distinguished German political journalist and writer, daughter of Theodor Boveri. Known to Warburg since 1913. Lived in Berlin-Dahlem after 1945.

Boveri, Theodor (1862–1915). German zoologist in Würzburg. Worked at the Zoological Station in Naples at the same time as Warburg. Founded the chromosome theory of heredity.

Bücher, Theodor (1914–). German biochemist, long-standing colleague of Warburg, later at Marburg and Munich.

Burk, Dean (1904–). US biochemist. Foreign Member of the Max Planck Institute for Cell Physiology since 1950.

Butenandt, Adolf (1903–). German biochemist, Danzig, Berlin, Tübingen and Munich. Nobel Prize 1939, for work on sex hormones. President of Max Planck Gesellschaft 1960–72.

Calvin, Melvin (1911–). US chemist and biochemist at Berkeley, California. Nobel Prize 1961 for work on photosynthesis ('The Calvin cycle').

Carnahan, James E, (1920–). US biologist at E. I. du Pont de Nemours & Co., Wilmington, Delaware.

Christian, Walter (1907–55). Long-standing technician and collaborator of Otto Warburg. Author and co-author of many publications.

Claude, Albert (1899–). Belgian–US medical student and biologist, Rockefeller Institute, New York; Universities of Brussels and Louvain. Nobel Prize 1974, for discoveries on the structural and functional organization of the cell.

Colowick, Sidney P. (1916–). US biochemist, Nashville, Tennessee.

Cori, Carl F. (1896–) and his wife, Gerty (1896–1957). US biochemists, mainly at Washington University, St. Louis, Missouri. Joint Nobel Prize 1947, for discoveries in the field of glycogen metabolism.

Correns, Carl (1864–1933). German plant-geneticist at the Kaiser Wilhelm Institute for Biology, Berlin-Dahlem. One of the rediscoverers of Mendel's Law.

Cremer, Werner. German co-worker with Warburg, around 1930.

Cuvier, Georges (1769–1832). French comparative anatomist and palaeontologist, Paris.

Davidson, J. Norman (1911–72). British biochemist. Visitor in Warburg's laboratory in 1938, later at London and Glasgow.

Davy, Sir Humphry (1778–1829). British chemist. From 1801 in London. Faraday's teacher. Among his many discoveries were the elements sodium and potassium. Inventor of the miner's safety lamp.

Embden, Gustav (1874–1933). German biochemist, Frankfurt am Main. Important contributions to the intermediate stages in lactic acid formation in muscle and liver metabolism.

Engelmann, Theodor Wilhelm (1843–1909). Physiologist at Utrecht and Berlin. Introduced Warburg to the field of photosynthesis.

Euler-Chepin, Hans von (1873–1964). German–Swedish biochemist, Stockholm. Nobel Prize 1929, for work on fermentation and enzymes of fermentation.

Fibiger, Johannes (1867–1928). Danish pathologist. Nobel Prize 1926, for work on the role of parasites in rat stomach tumours.

Fischer, Emil (1852–1919). One of the greatest German organic chemists, Berlin. One of Warburg's teachers. Nobel Prize 1902, for sugar and purine syntheses.

Fischer, Hans (1881–1945). German organic chemist, Munich. Nobel Prize 1930, for discoveries in the field of haemoglobin and chlorophyll.

Fleisch, Alfred (1892–1973). Swiss physiologist.

Franck, James (1882–1964). German–US physicist, Göttingen, Baltimore, Chicago. Nobel Prize 1925, for the discovery of the laws governing the impact of an electron upon an atom.

French, C. Stacy (1907–). US botanist, Stanford, California.

Gaffron, Hans (1902–79). German–US photobiologist. Warburg's colleague from 1925–30 and 1938–9. Later at Chicago and Tallahassee.

Gawehn, Karlfried (1927–). Long-standing colleague of Warburg. Later with the Boehringer Mannheim Company.

Goldblatt, Harry (1891–). US pathologist, Cleveland, Ohio.

Goldschmidt, Richard (1878–1958). German–US zoologist, leading geneticist, Berlin-Dahlem. Later (1936) Berkeley, California.

Haas, Erwin, (1904–77). Long-standing colleague of Warburg until 1938. Later at Chicago and Cleveland, Ohio.

Haber, Fritz (1868–1934). Leading German physical chemist, Karlsruhe and Berlin-Dahlem. Nobel Prize 1918 for his ammonia synthesis from its elements. (Basis for the Haber–Bosch process.)

Hahn, Otto (1879–1968). German physicist and chemist, Berlin-Dahlem. Nobel Prize 1944 for his discovery of the fission of heavy nuclei.

Haldane, John Scott (1860–1936). British physiologist. Main area of research, blood gases. Oxford and Birmingham.

Harden, Sir Arthur (1865–1940). British biochemist, London. Nobel Prize 1929 for discoveries in the field of alcoholic fermentation.

Hardie, Colin (1906–). British classical scholar, Oxford.

Harnack, Adolf von (1851–1930). German. Outstanding protestant theologian. Leipzig, Giessen, Marburg, Berlin. President (1910) of the Kaiser Wilhelm Gesellschaft, which had been founded on his suggestion.

Heilmeyer, Ludwig (1899–1969). German haematologist, Freiburg and Ulm.

Helmholtz, Hermann von (1821–94). Leading German physiologist and physicist, Königsberg, Bonn, Heidelberg, Berlin.

Henri, Victor (1872–1940). French physical chemist, Paris, Zürich, Liège.

Hess, Benno (1922–). German biochemist, Heidelberg, Dortmund.

Hilbert, David (1862–1943). Leading German mathematician. Göttingen.

Hill, Archibald Vivian (1886–1977). British physiologist, Cambridge, Manchester, London. Nobel Prize 1922 for work in the field of muscle physiology.

Hill, Robert (1899–). British biochemist, Cambridge. Work on photosynthesis.

Hoppe-Seyler, Ernst Felix Immanuel (1825–95). German physiologist and chemist, Strasbourg.

Hotchkiss, Rollin (1911–). US biochemist, Rockefeller University, New York.

Huxley, Sir Julian (1887–1975). British biologist. London.

Jaspers, Karl (1883–1969). German philosopher, Heidelberg, Basle.

Karrer, Paul (1889–1971). Swiss organic chemist, Zürich. Nobel Prize 1937 for work on carotinoids, flavins, and vitamins.

Keilin, David (1887–1963). British parasitologist and biochemist, Cambridge.

Kempner, Walter (1903–). Warburg's colleague in 1928 and from 1933–4. Later physician at Duke University, Durham, North Carolina.

Klein, Felix (1849–1928). Leading German mathematician, Göttingen.

Kleinzeller, Arnost (1914–). Czech–US biochemist and physiologist, Prague and Philadelphia.

Koch, Robert (1834–1910). Leading German bacteriologist, Berlin. Nobel Prize 1905 for the discovery of the tubercle bacillus.

Kraus, Friedrich (1858–1936). German, Professor of Internal Medicine, Berlin.

Krehl, Ludolf von (1861–1937). German, Professor of Internal Medicine, Heidelberg. Warburg's chief from 1906–14. Made advances in fundamental medical research.

Kroebel, Werner (1904–). German physicist, Kiel. Pupil of James Franck.

Kubowitz, Fritz (1902–). German, long-standing technical coworker of Warburg. Author and co-author of many publications. Later at Freiburg.

Kuhn, Richard (1900–67). Austrian–German organic chemist and biochemist, Zürich, Heidelberg. Nobel Prize 1938 for work on vitamins and carotinoids.

Kundt, August (1839–94). German physicist, Strasbourg, Berlin.

Kunitz, Moses (1887–1978). US biochemist. Rockefeller Institute, Princeton, and New York. Crystallization of proteins.

Laqueur, Ernst (1880–?). Professor of Pharmacology, University of Amsterdam.

Laue, Max von (1879–1960). German physicist. Pupil of Max Planck. Munich, Zürich, Frankfurt, Berlin, Göttingen. Nobel Prize 1914 for the discovery of the diffraction of X-rays by crystals.

Lehninger, Albert L. (1917–). US biochemist, Johns Hopkins University, Baltimore.

Leloir, Luis F. (1906–). Argentinian biochemist, Buenos Aires. Nobel Prize 1970 for the discovery of sugar nucleotides and their role in the biosynthesis of carbohydrates.

Lewis, Gilbert Newton (1875–1946). US physical chemist, Berkeley, California.

Lohmann, Karl (1898–1978). German biochemist. Meyerhof's colleague at Berlin-Dahlem. Later Professor at Berlin.

Lummer, Otto (1860–1923). German radiation physicist, Breslau.

Lynen, Feodor (1911–79). German biochemist, Munich. Nobel Prize 1964 for discoveries concerning mechanism and regulation of cholesterol and fatty acid metabolism.

MacMunn, Charles Alexander (1852–1911). Irish physician. Practised in Dublin and Wolverhampton. His research was carried out in his private laboratory.

Marwitz, Bodo von der (1893–). Gross-Kreutz, Cologne.

Maxwell, James Clark (1831–79). Leading British physicist, Cambridge and London.

Meyerhof, Otto (1884–1951). German biochemist, pupil of Warburg, Kiel, Heidelberg, Berlin-Dahlem, Philadelphia. Nobel Prize 1922 for work on the metabolism of muscle.

Mortenson, Leonard Earl (1928–). US biochemist with E. I. du Pont de Nemours, Wilmington and Purdue University, Lafayette, Indiana.

Myrbäck, Karl (1900–). Swedish biochemist.

Needham, Dorothy (1896–). British biochemist, Cambridge.

Negelein, Erwin (1897–1979). Long-standing German colleague of Warburg. Later Professor of Biochemistry in East Berlin.

Nernst, Walther (1864–1941). Leading German chemist and physicist. Göttingen and Berlin. Nobel Prize 1920 for his discovery of the third law of thermodynamics.

Neuberg, Carl (1877–1956). German biochemist. Berlin-Dahlem. Later at New York.

Niel, Cornelius van (1897–). US microbiologist (of Dutch extraction). Pacific Grove, California.

Ostwald, Wilhelm (1853–1932). German physical chemist, Riga, Leipzig. Nobel Prize 1909 for his work on catalysis and the kinetics of chemical reactions.

Parnas, Jacob (1884–1949). Polish biochemist, Lvov.

Planck, Max (1858–1947). German physicist, Kiel, Berlin. Discovered the energy quanta and thereby established the quantum theory. Nobel Prize 1918 (presented 1919).

Poincaré, Henri (1854–1912). Leading French mathematician, Paris.

Polanyi, Michael (1891–1976). German, of Hungarian extraction. Physician, physical chemist, economist, philosopher, Berlin-Dahlem, Manchester, Oxford.

Pringsheim, Ernst (1859–1917). German physicist (mainly radiation physics), Breslau.

Pullman, Maynard Edward (1927–). US biochemist, Johns Hopkins University, Baltimore, New York, Paris.

Racker, Efraim (1913–). US biochemist, Yale and Cornell Universities.

Rubner, Max (1854–1932). German physiologist, worked especially on energy metabolism, Munich, Marburg, Berlin.

Rutherford, Ernest (1871–1937). Leading British physicist, Montreal, Manchester, Cambridge. Nobel Prize 1908 for work on radioactivity.

Schmidt-Ott, Friedrich (1860–1956). German lawyer and historian; 1917–18 Prussian Minister of Education. In 1920, founded the Notgemeinschaft der Deutschen Wissenschaft and was its President until 1934. This was the forerunner of the Deutsche Forschungsrat zur Deutschen Forschungsgemeinschaft (the German equivalent of the British Research Councils and the US National Science Foundation and Institutes of Health).

Schneider, Walter C. (1919–). US biochemist, University of Wisconsin, Rockefeller Institute, New York, National Cancer Institute, Bethesda, Maryland.

Schoeller, Walther (1880–1965). German chemist, Director of the Central Research Laboratory of Schering AG, Berlin (1923–45) and Member of the Board of the Company. Longstanding friend of Warburg.

Slater, Edward Charles (1917–). Australian biochemist, Canberra, Cambridge, Amsterdam.

Smith, James Lorrain (1862–1931). British physiologist and pathologist, Belfast, Manchester, Edinburgh.

Szent-Györgyi, Albert von (1893–). Hungarian–US biochemist, Szeged, Budapest, Woods Hole, Massachusetts. Nobel Prize 1937 for discoveries in biological combustion processes, particularly in connection with Vitamin C.

Theorell, Hugo (1903–). Swedish biochemist, Stockholm. Worked in Warburg's laboratory, 1933–35. Nobel Prize 1955 for work on the enzymes of oxidation.

Thunberg, Torsten (1873–1952). Swedish physiologist, Lund.

Thomas, Karl (1883–1969). German physiological chemist, Leipzig, Erlangen, Göttingen.

Valentine, Raymond C. (1936–). US microbiologist and biochemist, University of California, San Diego.

Van Slyke, Donald D. (1883–1971). Leading US biochemist, Rockefeller Institute, New York.

Van't Hoff, Jacobus Henricus (1852–1911). Dutch chemist, Utrecht, Amsterdam, Berlin. Nobel Prize 1901 for work on stereochemistry and the theory of solutions.

Wallach, Otto (1847–1931). German organic chemist, Göttingen. Nobel Prize 1910 for pioneer investigations into the field of alicyclic compounds.

Weber, Hans Hermann (1896–1974). German physiologist, Rostock, Münster, Königsberg, Tübingen, Heidelberg.

Weber, Max (1864–1920). Leading German sociologist, Heidelberg, Munich.

Weinhouse, Sidney (1909–). US biochemist and cancer researcher, Philadelphia.

Whatley, Frederick Robert (1924–). British plant physiologist, Berkeley, California, London, Oxford.

Whittaker, Edmund (1873–1956). British mathematician, Edinburgh.

Wieland, Heinrich (1877–1957). German organic chemist, Freiburg, Munich. Nobel Prize 1927 for work on the constitution of bile-acid and related substances.

Wien, Wilhelm (1864–1928). German radiation physicist, Würzburg, Munich. Nobel Prize 1911 for the discovery of the laws governing the radiation of heat.

Willstätter, Richard (1872–1942). German chemist and biochemist, Zürich, Berlin-Dahlem, Munich. Nobel Prize 1915 for work with plant pigments, especially chlorophyll.

Wren, Sir Christopher (1623–1723). British mathematician, astronomer and architect, London and Oxford.

References

(1) Andrade, E. N. da C. (1964). Galileo. *Notes Rec. R. Soc. Lond.* **19,** 120–30.

(2) Arnon, D. I. (1969). Role of ferredoxin in photosynthesis. *Naturwissenschaften,* **56,** 295–305.

(3) Arnon, D. I. (1971). The light reactions of photosynthesis. *Proc. natn. Acad. Sci. USA.* **68,** 2883–92.

(4) Bangham, A. D. (1972). Lipid bilayers and biomembranes. *A. Rev. Biochem.* **41,** 753–76.

(5) Barron, E. S. G. and Harrop, G. A. Jr. (1928). Studies on blood cell metabolism. *J. biol. Chem.* **79,** 65–87.

(6) Bayliss, W. M. (1920). *Principles of general physiology.* Longman Green, London.

(7) Beevers, L. and Hageman, R. H. (1969). Nitrate reduction in higher plants. *A. Rev. Pl. Physiol.* **20,** 495–522.

(8) Broyer, T. C., Carlton, A. B., Johnson, C. M., and Stout, P. R. (1954). Chlorine—a micro-nutrient element for higher plants. *Pl. Physiol., Lancaster* **29,** 526–32.

(9) Burke, P. (1971). The Warburg tradition. *The Listener,* 21 October, 546–8. (An appreciation of Aby Warburg.)

(10) Candau, P., Manzano, C., and Losada, M. (1976). Bioconversion of light energy into chemical energy through reduction with water of nitrate to ammonia. *Nature, Lond.* **262,** 715–17.

(11) Cheniae, G. M. (1970). Photosystems II and O_2 evolution. *A. Rev. Pl. Physiol.* **21,** 467–98.

(12) Chorine, V. (1945). Action de l'amide nicotinique sur les bacilles de genre Mycobacterium. *C. r. hebd. Séanc. Acad. Sci., Paris,* **220,** 150–1.

(13) Claude, A. (1947–8). Studies on cells. Morphology, chemical constitution, and distribution of biochemical functions. *Harvey Lect.* **43,** 121–64.

(14) Cuvier, G. L. (1829). *Éloge. Mém. Acad. R. Sci. l'Instit. France,* **12,** LVI.

(15) Davenport, H. E., Hill, R., and Whatley, F. R. (1952). A natural factor catalysing reduction of methaemoglobin by isolated chloroplasts. *Proc. R. Soc.* **B139,** 346–58.

(16) Earle, W. R. and Nettleship, A. (1943). Production of malignancy *in vitro*. V. Results of injections of cultures into mice. *J. natn. Cancer Inst.* **4**, 247–9.
(17) Einstein–Born correspondence (1916–55). Nymphenburger Verlagshandlung GmbH, Munich (1969).
(18) Elvehjem, C. A., Madden, R. J., Strong, F. M., and Woolley, D. W. (1937). Relation of nicotinic acid and nictonic acid amide to canine black tongue. *J. A. chem. Soc.* **59**, 1767.
(19) Evans, V. J. and Andersen, W. F. (1966). Effect of serum on spontaneous neoplastic transformations *in vitro*. *J. natn. Cancer Inst.* **37**, 247–9.
(20) Evans, V. J., Parker, G. A., and Dunn, Thelma B. (1964). Neoplastic transformation in C3H mouse embryonic tissue *in vitro* determined by intraocular growth. 1. Cells from chemically defined medium with and without serum supplement. *J. natn. Cancer Inst.* **32**, 89–107.
(21) Fick, A. (1893). *Mechanische Arbeit und Wärmeentwicklung bei der Muskeltätigkeit.* Verlag Brockhaus, Leipzig.
(22) Franck, J. (1931). Emil Warburg zum Gedächtnis. *Naturwissenschaften.* **19**, 993–7.
(23) Gey, G. O. (1941). Cytological and cultural observations on transplantable rat sarcomata produced by the inoculation of altered normal cells maintained in continuous culture. *Cancer Res.* **1**, 737.
(24) Goethe, J. W. v. *Faust,* Part 2, Act 2 line 6771.
(25) Goethe, J. W. v. *Faust,* Part 2, Act 2, lines 11936–7.
(26) Goethe, J. W. v. (1972). *Der Gross-Cophta,* Scene 4.
(27) Goldblatt, H. and Cameron, G. (1953). Induced malignancy in cells from rat myocardium subjected to intermittent anaerobiosis during long propagation *in vitro*. *J. exp. Med.* **97**, 525–52.
(28) Haldane, J. S. and Smith, J. L. (1896). The oxygen tension in arterial blood. *J. Physiol., Lond.* **20**, 497–520.
(29) Heilmeyer, L. and Plötner, K. (1937). *Das Serumeisen und die Eisenmangelkrankheit.* Jena.
(30) Heilmeyer, L. and Strüwe, G. (1938). Der Eisen-Kupfer-antagonismus im Blutplasma beim Infektionsgeschehen. *Klin. Wschr.* **17**, 925–7.
(31) Henri, V. (1926). Die spezifische photochemische Wirkung bei der Kohlensäureassimilation nach den Versuchen von Wurmser. *Naturwissenschaften* **14**, 165–7.
(32) Hertz, H. A. (1937). *Stamm- und Nachfahrentafel der Familie Warburg*, Hamburg-Altona, Hamburg. [Published for the family.]
(33) Hess, B. (1962). *Enzyme im Blutplasma.* Georg Thieme Verlag,

Stuttgart. [*Enzymes in blood plasma* (translated by K. S. Hensley). Academic Press, New York (1963).]

(34) Hill. R. (1965). The photosynthetic electrontransport in plants. *Essays Biochem.* **1**, 121–51.

(35) Izawa, S., Heath, R. L., and Hind, G. (1969). The role of chloride ion in photosynthesis. *Biochim. biophys. Acta*, **180**, 388–98.

(36) Keilin, D. (1966). *The history of cell respiration and of cytochrome*. Cambridge University Press.

(37) Keilin, D. and Hartree, E. F. (1939). Cytochrome and cytochrome oxidase. *Proc. R. Soc.* **B127**, 167–91.

(38) Keilin, D. (1929). Cytochrome and respiratory enzymes. *Proc. R. Soc.* **B104**, 205–52.

(39) Kent, G. O. (1968). *Arnim and Bismarck*. Oxford University Press.

(40) Kleinzeller, A. (ed.) (1965). *Manometrische Methoden*. Gustav Fischer Verlag, Jena.

(41) Krebs, H. A. (1972). The Pasteur effect and the relations between respiration and fermentation. *Essays Biochem.* **8**, 1–34.

(42) Kroebel, W. (1964). Zum Tode von James Franck. *Naturwissenschaften*, **51**, 421–3.

(43) *The Lancet* (Editorial) (1950). *Lancet*, **ii**, 688.

(44) Leloir, E. F. (1971). Two decades of research on the biosynthesis of saccharides. *Science, N.Y.* **172**, 1299–303.

(45) Lewis, G. N. and Randall, M. (1923). *Thermodynamics and the free energy of chemical substances*. McGraw-Hill, New York.

(46) Liljestrand, G. (1950). The prize in physiology and medicine. In, *Nobel, the man and his prizes*. Sohlmans Förlag, Stockholm.

(47) Losada, M., Ramirez, J. M., Paneque, A., and Del Campo, F. F. (1965). Light and dark reduction of nitrate in a reconstituted chloroplast system. *Biochim. biophys. Acta*, **109**, 86–96.

(48) Losada, M. (1976). Reducing power and the regulation of photosynthesis. In *Reflections on biochemistry*, (ed. B. L. Horecker, L. Cordudella, and J. Oro), pp. 73–83. Pergamon Press, Oxford.

(49) Lovenberg, W. (ed.), (1973, 1974, 1977). *Iron-sulphur proteins*, 3 Vols. Academic Press, New York.

(50) Mann, T. (1964). David Keilin. *Biogr. Mem. Fellows R. Soc.* **10**, 183–205.

(51) Martin, G. and Lavolley, J. (1958). Le chlore, oligo-élément indispensable pour Lemna minor. *Experientia*, **14**, 333.

(52) Marwedel, G. (1976). Die Privilegien der Juden in Altona. Vol. 5 of the *Beiträge zur Geschichte der deutschen Juden*. Hans Christian Verlag, Hamburg.

(53) Maxwell, J. C. (1883). *Lehrbuch der Electricität und des Magnetismus*. Springer-Verlag, Berlin.

(54) Mortenson, L. E., Valentine, R. C., and Carnahan, J. E. (1962). An electron transport factor from Clostridium pasteurianum. *Biochem. biophys. Res. Commun.* **7**, 448–52.

(55) Niel, C. B. van. (1962). The present status of the comparative study of photosynthesis. *A. Rev. Pl. Physiol.* **13**, 1–26.

(56) Orme-Johnson, W. H. (1973). Iron-sulphur proteins, structure and function. *A. Rev. Biochem.* **42**, 159–98.

(57) Panofsky, E. (1930). In a privately printed work, *Nachrufe auf Aby Warburg*.

(58) Planck, M. (1948). *Wissenschaftliche Selbstbiographie*, p. 22. Johann Ambrosius Barth-Verlag, Leipzig.

(59) Pullman, M. E., San Pietro, A., and Colowick, S. P. (1954). On the structure of reduced diphosphopyridine nucleotide. *J. biol. Chem.* **206**, 129–41

(60) Racker, E. (1972). Bioenergetics and the problem of tumour growth. *Am. Scient.* **60**, 56–63.

(61) Reichert, I. (1938). Otto Warburg. *Palest. J. Bot. Rehovot Ser.* **2**, 2–16. [An appreciation of the botanist Otto Warburg.]

(62) Robertson, N. L. and Boardman, N. K. (1975). The link between charge separation, proton movement and ATPase reactions. *FEBS Letts.* **60**, 1–6.

(63) Rosenbaum, E. and Sherman, A. J. (1976). *Das Bankhaus M. M. Warburg & Co. 1798–1938*. Hans Christian Verlag, Hamburg.

(64) Sibley, J. A. and Lehninger, A. L. (1949). *J. biol. Chem.* **177**, 859–72.

(65) Slater, E. C. (1964). In memoriam David Keilin. *Enzymologia*, **26**, 313–20.

(66) Stern, K. G. and Holiday, E. R. (1934). Die Photoflavine, eine Gruppe von Alloxazin-Derivaten. *Ber. dt. chem. Ges.* **67**, 1442–52.

(67) Theorell, H. (1970). In *Pyridine nucleotide-dependent dehydrogenases* (ed. H. Sund). Springer-Verlag, Berlin.

(68) Warburg, E. (1930). In a privately printed volume, *Aby M. Warburg zum Gedächtnis*.

(69) Warburg, F. S. (1914). *Die Geschichte der Firma R. D. Warburg, ihre Teilhaber und deren Familien*. Privatdruck Otto Dreyer, Berlin W. 57, Berlin. [Obtainable in: Deutsche Staatsbibliothek, DDR-108 Berlin, Postfach 1312.]

(70) Wechsberg, J. (1966). Profile: a prince of the city. *New Yorker*, 9 April, 45–77. [A 'profile' of Siegmund Warburg.]

(71) Weinhouse, S. (1956). Oxidative metabolism of neoplastic tissue. *Science N.Y.* **124**, 267–9.

(72) Willstätter, R. (1926). Über Fortschritte in der Enzym-Isolierung. *Ber. dt. chem. Ges.* **59**, 1–12.
(73) Willstätter, R. (1958). *Aus meinem Leben* (2nd edn). Verlag Chemie, S. 200, Weinheim.

Bibliography

The list of publications contains a number of papers from Warburg's laboratory published without his name. These papers were included because Warburg himself was a major contributor to the work. He was generous in giving credit to his juniors.

The bibliography, though extensive, is not complete. It was the intention to include all essential papers.

[1] 1905. Spaltung des Leucinesters durch Pankreasferment. *Ber. dt. chem. Ges.* **38**, 187.

[2] 1905. (With E. FISCHER.) Spaltung des Leucins in die optisch-aktiven Komponenten mittels der Formylverbindung. *Ber. dt. chem. Ges.* **38**, 3997.

[3] 1905. (With E. FISCHER.) Synthese von Polypeptiden. Glycl-leucin, Alanyl-leucin, Leucyl-alanin, Glycl-alanyl-leucin und aktives Alanyl-glycin. *Justus Liebigs Annln. Chem.* **340**, 152.

[4] 1905. (With E. FISCHER.) Optisch aktive α-Brom propionsäure. *Justus Liebigs Annln. Chem.* **340**, 168.

[5] 1906. Spaltung des Leucinesters durch Pankreasferment. *Hoppe-Seyler's Z. physiol. Chem.* **48**, 205.

[6] 1908. Beobachtungen über die Oxydationsprozesse im Seeigelei. *Hoppe-Seyler's Z. physiol. Chem.* **57**, 1.

[7] 1909. Zur Biologie der roten Blutzellen. *Hoppe-Seyler's Z. physiol. Chem.* **59**, 112.

[8] 1909. Über die Oxydationen im Ei. *Hoppe-Seyler's Z. physiol. Chem.* **60**, 443.

[9] 1909. Massanalytische Bestimmung kleiner Kohlensäuremengen. *Hoppe-Seyler's Z. physiol Chem.* **61**, 261.

[10] 1910. Über die Oxydationen in lebenden Zellen nach Versuchen am Seeigelei. *Hoppe-Seyler's Z. physiol. Chem.* **66**, 305.

[11] 1910. Über Beeinflussung der Oxydationen in lebenden Zellen nach Versuchen an roten Blutkörperchen. *Hoppe-Seyler's Z. physiol. Chem.* **69**, 452.

[12] 1910. Bemerkung zu einer Arbeit von Jacques Loeb und Wasteneys. *Hoppe-Seyler's Z. physiol. Chem.* **69**, 496.

[13] 1910. (With O. Neubauer.) Über eine Synthese mit Essigsäure in der künstlich durchbluteten Leber. *Hoppe-Seyler's Z. physiol. Chem.* **70**, 1.

[14] 1910. Über die giftige Wirkung der Natriumchloridlösung. *Bioch. Z.* **29**, 414.

[15] 1911. Über Beeinflussung der Sauerstoffatmung. *Hoppe-Seyler's Z. physiol. Chem.* **70**, 413.

[16] 1911. (M. Onaka.) Über die Wirkung des Arsens auf die roten Blutzellen. *Hoppe-Seyler's Z. physiol. Chem.* **70**, 433.

[17] 1911. (M. Onaka.) Über Oxydationen im Blut. *Hoppe-Seyler's Z. physiol. Chem.* **71**, 193.

[18] 1911. Über Beeinflussung der Sauerstoffatmung. *Hoppe-Seyler's Z. physiol. Chem.* **71**, 479.

[19] 1911. Beziehungen zwischen Konstitution und physiologischer Wirkung. *Verh. d. Dtsch. Kongr. f. innere Med.* **28**, 553.

[20] 1911. Untersuchungen über die Oxydationsprozesse in Zellen. *Münch. med. Wschr.* **58**, 289.

[21] 1912. Untersuchungen über die Oxydationsprozesse in Zellen (II). *Münch. med. Wschr.* **59**, 2550.

[22] 1912. Über Hemmungen der Blausäurewirkung in lebenden Zellen. *Hoppe-Seyler's Z. physiol. Chem.* **76**, 331.

[23] 1912. Notiz über Bestimmung kleiner, im Wasser gelöster CO_2-Mengen. *Hoppe-Seyler's Z. physiol. Chem.* **81**, 202.

[24] 1912. (R. Usui.) Über die Bindung von Thymol in roten Blutzellen. *Hoppe-Seyler's Z. physiol. Chem.* **81**, 175.

[25] 1912. (With R. Wiesel.) Über die Wirkung von Substanzen homologer Reihen auf Lebensvorgänge. *Pflüger's Arch. ges. Physiol.* **144**, 465.

[26] 1912. Über Beziehungen zwischen Zellstruktur und biochemischen Reaktionen. I. *Pflüger's Arch. ges. Physiol.* **145**, 277.

[27] 1912. (With O. Meyerhof.) Über Atmung in abgetöteten Zellen und in Zellfragmenten. *Pflüger's Arch. ges. Physiol.* **148**, 295.

[28] 1912. (E. Grafe.) Über die Wirkung von Ammoniak und Ammoniakderivaten auf die Oxydationsprozesse in Zellen. *Hoppe-Seyler's Z. physiol. Chem.* **79**, 421.

[29] 1912. (A. Dorner.) Über Beeinflussung der alkoholischen

Gärung in der Zelle und im Zellpressaft. *Hoppe-Seyler's Z. physiol. Chem.* **81**, 99.

[30] 1912. (R. WIESEL.) Über die Wirkung von Blutserum auf die Oxydationsprozesse von Bakterien. *Z.f.S. Immunitätsforch.* **12**, 194.

[31] 1912. (R. USUI.) Über Messung von Gewebsoxydationen *in vitro. Pflüger's Arch. ges. Physiol.* **147**, 100.

[32] 1913. (A. DORNER.) Über Titration kleiner Kohlensäuremengen. *Hoppe-Seyler's Z. physiol. Chem.* **88**, 425.

[33] 1913. Über sauerstoffatmende Körnchen aus Leberzellen und über Sauerstoffatmung in Berkefeld-Filtraten wässriger Leberextrakte. *Pflüger's Arch. ges. Physiol.* **154**, 599.

[34] 1913. Antwort auf vorstehende Bemerkung Thunbergs. *Hoppe-Seyler's Z. physiol. Chem.* **87**, 83.

[35] 1914. Über die Rolle des Eisens in der Atmung des Seeigeleis nebst Bemerkungen über einige durch Eisen beschleunigten Oxydationen. *Hoppe-Seyler's Z. physiol. Chem.* **92**, 231.

[36] 1914. Über Verbrennung der Oxalsäure an Blutkohle und die Hemmung dieser Reaktion durch indifferente Narkotika. *Pflüger's Arch. ges. Physiol.* **155**, 547.

[37] 1914. Über die Empfindlichkeit der Sauerstoffatmung gegenüber indifferenten Narkotika. *Pflüger's Arch. ges. Physiol.* **158**, 19.

[38] 1914. Zellstruktur und Oxydationsgeschwindigkeit nach Versuchen am Seeigelei. *Pflüger's Arch. ges. Physiol.* **158**, 189.

[39] 1914. (A. DORNER.) Über Verteilungsgleichgewichte einiger indifferenter Narkotika *Sitzungsberichte Heidelberg. Akad. Wiss. Math.-Nat. Klasse* Abt. B. **1**. Abhandlung, 3.

[40] 1914. Beiträge Physiologie der Zelle, insbesondere über die Oxydationsgeschwindigkeit in Zellen. *Ergebn. Physiol.* **14**, 253.

[41] 1915. Notizen zur Entwicklungsphysiologie des Seeigeleis. *Pflüger's Arch. ges. Physiol.* **160**, 324.

[43] 1919. Über die Geschwindigkeit der photochemischen Kohlensäurezersetzung in lebenden Zellen. *Biochem. Z.* **100**, 230.

[44] 1920. Über die Geschwindigkeit der photochemischen Kohlensäurezersetzung in lebenden Zellen II. *Biochem. Z.* **103**, 188.

[45] 1920. (With E. NEGELEIN.) Über die Reduktion der Salpetersäure in grünen Zellen. *Biochem. Z.* **110**, 66.

[46] 1921. (With E. NEGELEIN.) Über die Oxydation des Cystins und anderer Aminosäuren an Blutkohle. *Biochem. Z.* **113**, 257.

[47] 1921. Physikalische Chemie der Zellatmung. *Biochem. Z.* **119**, 134.

[48] 1921. Theorie der Kohlensäureassimilation. *Naturwissenschaften* **9**, 354.

[49] 1922. (With E. NEGELEIN.) Über den Energieumsatz bei der Kohlensäureassimilation. *Z. Electrochem.* **28**, 449

[50] 1922. Über Oberflächenreaktionen in lebenden Zellen. *Z. Electrochem.* **28**, 70.

[51] 1922. (With E. NEGELEIN.) Über den Energieumsatz bei der Kohlensäureassimilation. *Z. phys. Chem.* **102**, 235.

[52] 1923. (With E. NEGELEIN.) Über den Einfluss der Wellenlänge auf den Energieumsatz bei der Kohlensäureassimilation. *Z. Phys. Chem.* **106**, 191.

[53] 1923. (With E. NEGELEIN.) Bemerkung zu einem Aufsatz von F. Weigert. *Z. Phys. Chem.* **108**, 101.

[54] 1923. (With S. SAKUMA.) Über die sogenannte Autoxydation des Cysteins. *Pflüger's Arch. ges. Physiol.* **200**, 203.

[55] 1923. Über die antikatalytische Wirkung der Blausäure. *Biochem. Z.* **136**, 266.

[56] 1923. (S. SAKUMA.) Über die sogenannte Autoxydation des Cysteins. *Biochem. Z.* **142**, 68.

[57] 1923. Versuche an überlebendem Carcinomgewebe. *Biochem. Z.* **142**, 317.

[58] 1923. (S. MINAMI.) Versuche an überlebendem Carcinomgewebe. *Biochem. Z.* **142**, 334.

[59] 1923. (E. NEGELEIN.) Über die Reaktionsfähigkeit verschiedener Aminosäuren an Blutkohle. *Biochem. Z.* **142**, 493.

[60] 1923. Über die Grundlagen der Wielandschen Atmungstheorie. *Biochem. Z.* **142**, 518.

[61] 1924. (With W. BREFELD.) Über die Aktivierung stickstoffhaltiger Kohlen durch Eisen. *Biochem. Z.* **145**, 461.

[62] 1924. (With M. YABUSOE.) Über die Oxydation von Fructose in Phosphatlösungen. *Biochem. Z.* **146**, 380.

[63] 1924. (With T. UYESUGI.) Über die Blackmansche Reaktion. *Biochem. Z.* **146**, 486.

[64] 1924. Verbesserte Methode zur Messung der Atmung und Glykolyse. *Biochem. Z.* **152**, 51.

[65] 1924. Bemerkung über das Kohlemodell. *Biochem. Z.* **152,** 191.

[66] 1924. (With K. POSENER and E. NEGELEIN.) Über den Stoffwechsel der Carcinomzelle. *Biochem. Z.* **152,** 309.

[67] 1924. Über Eisen, den sauerstoffübertragenden Bestandteil des Atmungsferments. *Biochem. Z.* **152,** 479.

[68] 1924. (M. YABUSOE.) Über den Temperaturkoeffizienten der Kohlensäureassimilation. *Biochem. Z.* **152,** 498.

[69] 1925. (M. YABUSOE.) Über Eisen- und Blutfarbstoffbestimmungen in normalen Geweben und in Tumorgewebe. *Biochem. Z.* **157,** 388.

[70] 1925. (K. TANAKA.) Versuche zur Prüfung der Wielandschen Atmungstheorie. *Biochem. Z.* **157,** 425.

[71] 1925. (E. NEGELEIN.) Versuche über Glykolyse. *Biochem. Z.* **158,** 121.

[72] 1925. (F. WIND.) Über die Oxydation von Dioxyaceton und Glycerinaldehyd in Phosphatlösungen und die Beschleunigung der Oxydation durch Schwermetalle. *Biochem. Z.* **159,** 58.

[73] 1925. (Y. OKAMOTO.) Über Anaerobiose von Tumorgewebe. *Biochem. Z.* **160,** 52.

[74] 1925. Über Milchsäurebildung beim Wachstum. *Biochem. Z.* **160,** 307.

[75] 1925. Bemerkung zu einer Arbeit von M. Dixon and S. Thurlow sowie zu einer Arbeit von G. Ahlgren. *Biochem. Z.* **163,** 252.

[76] 1925. (E. NEGELEIN.) Über die Wirkung des Schwefelwasserstoffs auf chemische Vorgänge in Zellen. *Biochem. Z.* **165,** 203.

[77] 1925. Manometrische Messung des Zellstoffwechsels in Serum. *Biochem. Z.* **164,** 481.

[78] 1925. (E. NEGELEIN.) Über die glykolytische Wirkung des embryonalen Gewebes. *Biochem. Z.* **165,** 122.

[79] 1925. Über die Wirkung der Blausäure auf die alkoholische Gärung. *Biochem. Z.* **165,** 196.

[80] 1925. Versuche über die Assimilation der Kohlensäure. *Biochem. Z.* **166,** 386.

[81] 1925. (H. GAFFRON.) Über eine photochemische Wirkung des Hämatoporphyrins. *Naturwissenschaften* **13,** 859.

[82] 1925. Über Eisen, den sauerstoffübertragenden Bestandteil des Atmungsferments. *Chem. Ber.* **58,** 1001.

[83] 1926. (M. YABUSOE.) Über Hemmung der Tumorglykolyse durch Anilinfarbstoffe . *Biochem. Z.* **168,** 227.

[84] 1926. (S. TODA.) Über die Oxydation der Oxalsäure durch Jodsäure in wässeriger Lösung. *Biochem. Z.* **171**, 231.

[85] 1926. (S. TODA.) Über die Wirkung von Blausäureäthylester (Äthylcarbylamin) auf Schwermetallkatalysen. *Biochem. Z.* **172**, 17.

[86] 1926. (S. TODA.) Über "Wasserstoffaktivierung" durch Eisen. *Biochem. Z.* **172**, 34.

[87] 1926. Über die Wirkung von Blausäureäthylester (Äthylcarbylamin) auf die Pasteursche Reaktion. *Biochem. Z.* **172**, 432.

[88] 1926. Über die Oxydation der Oxalsäure durch Jodsäure. *Biochem. Z.* **174**, 497.

[89] 1926. Über die Wirkung des Kohlenoxyds auf den Stoffwechsel der Hefe. *Biochem. Z.* **177**, 471.

[90] 1926. (F. WIND.) Versuche über den Stoffwechsel von Gewebsexplantaten und deren Wachstum bei Sauerstoff- und Glucosemangel. *Biochem. Z.* **179**, 384.

[91] 1926. (H. GAFFRON.) Über Photoxydationen mittels fluoreszierender Farbstoffe. *Biochem. Z.* **179**, 157.

[92] 1926. (H. A. KREBS.) Über die Rolle der Schwermetalle bei der Autoxydation von Zuckerlösungen. *Biochem. Z.* **180**, 377.

[93] 1926. Atmungstheorie und Katalase. *Chem. Ber.* **59**, 739.

[94] 1926. Über die Wirkung von Kohlenoxyd und Licht auf den Stoffwechsel der Hefe. *Naturwissenschaften* **14**, 471.

[95] 1926. Erwiderung auf einen Aufsatz von Henri. *Naturwissenschaften* **14**, 167.

[96] 1926. (E. NEGELEIN.)—Über Abtötung von Tumorzellen im Körper. *Naturwissenschaften* **14**, 439.

[97] 1926. Über künstliche Mineralwässer. *Klin. Wschr.* **5**, 734.

[98] 1926. (With F. WIND and E. NEGELEIN.) Über den Stoffwechsel von Tumoren im Körper. *Klin. Wschr.* **5**, 829.

[99] 1926. (With O. STAHL.) Über Milchsäuregärung eines menschlichen Blasencarcinoms. *Klin. Wschr.* **5**, 1218.

[100] 1926. (F. WIND.) Versuche mit explantiertem Roussarkom. *Klin. Wschr.* **5**, 1355.

[101] 1926. Über Carcinomversuche. *Klin. Wschr.* **5**, 2119.

[102] 1926. Das Carcinomproblem. *Z. Angew. Chem.* **39**, 949.

[103] 1927. Über den heutigen Stand des Carcinomproblems. *Naturwissenschaften* **15**, 1.

[104] 1927. (With F. WIND and E. NEGELEIN.) The metabolism of tumors in the body. *J. gen. Physiol.* **8**, 519.

[105] 1927. (W. FLEISCHMANN and F. KUBOWITZ.) Über den Stoffwechsel der Leucocyten. *Biochem. Z.* **181**, 395.

[106] 1927. Über die Klassifizierung tierischer Gewebe nach ihrem Stoffwechsel. *Biochem. Z.* **184**, 484.

[107] 1927. Über den Stoffwechsel der Leucocyten. *Biochem. Z.* **187**, 1.

[108] 1927. Methode zur Bestimmung von Kupfer und Eisen und über den Kupfergehalt des Blutserums. *Biochem. Z.* **187**, 255.

[109] 1927. (H. A. KREBS.) Über den Stoffwechsel der Netzhaut. *Biochem. Z.* **189**, 57.

[110] 1927. (C. TAMIYA.) Über den Stoffwechsel der Netzhaut in verschiedenen Stadien ihrer Entwicklung. *Biochem. Z.* **189**, 114.

[111] 1927. (C. TAMIYA.) Über den Stoffwechsel der Leber in verschiedenen Stadien ihrer Entwicklung. *Biochem. Z.* **189**, 175.

[112] 1927. (H. A. KREBS and F. KUBOWITZ.) Über den Stoffwechsel von Carcinomzellen in Carcinomserum und Normalserum. *Biochem. Z.* **189**, 194.

[113] 1927. (With F. KUBOWITZ.) Stoffwechsel wachsender Zellen (Fibroblasten, Herz, Chorion). *Biochem. Z.* **189**, 242.

[114] 1927. Über den Stoffwechsel der Hefe. *Biochem. Z.* **189**, 350.

[115] 1927. Über die Wirkung von Kohlenoxyd und Stickoxyd auf Atmung and Gärung. *Biochem. Z.* **189**, 354.

[116] 1927. (With H. A. KREBS.) Über locker gebundenes Kupfer und Eisen im Blutserum. *Biochem. Z.* **190**, 143.

[117] 1927. (G. ENDRES and F. KUBOWITZ.) Stoffwechsel der Blutplättchen. *Biochem. Z.* **191**, 395.

[118] 1927. Über Kupfer im Blutserum des Menschen. *Klin. Wschr.* **6**, 109.

[119] 1927. (A. FUJITA.) Über den Stoffwechsel der weissen Blutzellen. *Klin. Wschr.* **7**, 897.

[120] 1927. Milchsäuregärung der Tumoren. *Klin. Wschr.* **6**, 2047.

[121] 1927. (W. KEMPNER.) Atmung im Plasma Pestkranker Hühner. *Klin. Wschr.* **6**, 2386.

[122] 1927. Über reversible Hemmung von Gärungsvorgängen durch Stickoxyd. *Naturwissenschaften* **15**, 51.

[123] 1927. Über Kohlenoxydwirkung ohne Hämoglobin und einige Eigenschaften des Atmungsferments. *Naturwissenschaften* **15**, 546.

[124] 1927. (H. GAFFRON.) Sauerstoffübertragung durch Chloro-

phyll und das photochemische Äquivalent-Gesetz. *Chem. Ber.* **60**, 755.

[125] 1927. (H. Gaffron.) Die photochemische Bildung von Peroxyd bei der Sauerstoff-Übertragung durch Chlorophyll. *Chem. Ber.* **60**, 2229.

[126] 1928. (W. Cremer.) Notiz über die Reduktion von Hämin durch Cystein. *Biochem. Z.* **192**, 426.

[127] 1928. (S. Kumanomido.) Stoffwechsel embryonaler Gewebe in Serum. *Biochem. Z.* **193**, 315.

[128] 1928. (With E. Negelein.) Über die Verteilung des Atmungsferments zwischen Kohlenoxyd und Sauerstoff. *Biochem. Z.* **193**, 334.

[129] 1928. (With E. Negelein.) Über den Einfluss der Wellenlänge auf die Verteilung des Atmungsferments. *Biochem. Z.* **193**, 339.

[130] 1928. (H. A. Krebs.) Über die Wirkung von Kohlenoxyd und Licht auf Häminkatalysen. *Biochem. Z.* **193**, 347.

[131] 1928. (W. Cremer.) Über eine Kohlenoxydverbindung des Ferrocysteins und ihre Spaltung durch Licht. *Biochem. Z.* **194**, 231.

[132] 1928. (F. Wind and K. v. Oettingen.) Milchsäurebestimmung in den Uterus- und Nabelgefässen, *Biochem. Z.* **197**, 170.

[133] 1928. (A. Fujita.) Über den Stoffwechsel der Körperzellen. *Biochem. Z.* **197**, 175.

[134] 1928. (A. Fujita.) Über die Wirkung des Kohlenoxyds auf den Stoffwechsel der weissen Blutzellen. *Biochem. Z.* **197**, 189.

[135] 1928. Über die oxydationskatalytische Wirkung des Eisens nach Handovsky. *Biochem. Z.* **198**, 241.

[136] 1928. (With E. Negelein.) Über die photochemische Dissoziation von Eisencarbonylverbindungen (Kohlenoxyd-Hämochromogen, Kohlenoxyd-Ferrocystein) und das photochemische Äquivalentgesetz. *Biochem. Z.* **200**, 414.

[137] 1928. (H. A. Krebs.) Über die Wirkung des Kohlenoxyds auf Hämatinkatalysen nach M. Dixon. *Biochem. Z.* **201**, 489.

[138] 1928. (W. Cremer.) Über Hemmung der Ferrocysteinkatalyse durch Kohlenoxyd. *Biochem. Z.* **201**, 490.

[139] 1928. Wie viele Atmungsfermente gibt es? *Biochem. Z.* **201**, 486.

[140] 1928. (With E. Negelein.) Über die photochemische Dis-

soziation bei intermittierender Belichtung und das absolute Absorptionsspektrum des Atmungsferments. *Biochem. Z.* **202**, 202.

[141] 1928. (With F. KUBOWITZ.) Notiz über manometrische Messung kleiner Sauerstoffpartialdrucke. *Biochem. Z.* **202**, 387.

[142] 1928. (With F. KUBOWITZ.) Über die Konzentration des Fermenteisens in der Zelle. *Biochem. Z.* **203**, 95.

[143] 1928. (M. NAKASHIMA.) Stoffwechsel der Fischnetzhaut bei verschiedenen Temperaturen. *Biochem. Z.* **204**, 479.

[144] 1928. Über die chemische Konstitution des Atmungsfermentes. *Naturwissenschaften* **16**, 345.

[145] 1928. (With E. NEGELEIN.) Über die photochemische Spaltung einer Eisencarbonylverbindung und das photochemische Äquivalentgesetz. *Naturwissenschaften* **16**, 387.

[146] 1928. Photochemie der Eisencarbonylverbindungen und das absolute Absorptionsspecktrum des Atmungsferments. *Naturwissenschaften* **16**, 856.

[147] 1928. Stoffwechsel der Karzinomelle. *Verdlg. dtsch. Kongr. f. innere Med.* **40**, 11.

[148] 1929. (With E. NEGELEIN.) Absolutes Absorptionsspektrum des Atmungsferments. *Biochem. Z.* **204**, 495.

[149] 1929. (H. A. KREBS.) Über die Wirkung von Kohlenoxyd und Blausäure auf Hämatinkatalysen. *Biochem. Z.* **204**, 322.

[150] 1929. (H. A. KREBS.) Über die Wirkung der Schwermetalle auf die Autoxydation der Alkalisulfide und des Schwefelwasserstoffs. *Biochem. Z.* **204**, 343.

[151] 1929. (F. KUBOWITZ.) Stoffwechsel der Froschnetzhaut bei verschiedenen Temperaturen und Bemerkung über den Meyerhofquotienten bei verschiedenen Temperaturen. *Biochem. Z.* **204**, 475.

[152] 1929. Ist die aerobe Glykolyse spezifisch für die Tumoren? *Biochem. Z.* **204**, 482.

[153] 1929. (W. CREMER.) Reaktionen des Kohlenoxyds mit Metallverbindungen des Cysteins. *Biochem. Z.* **206**, 228.

[154] 1929. Bemerkung zu einer Arbeit von D. Keilin. *Biochem. Z.* **207**, 494.

[155] 1929. (H. A. KREBS.) Über Hemmung einer Hämatinkatalyse durch Schwefelwasserstoff. *Biochem. Z.* **209**, 32.

[156] 1929. (H. A. Krebs and J. F. Donegan.) Manometrische Messung der Peptidspaltung. *Biochem. Z.* **210**, 7.

[157] 1929. Atmungsferment und Oxydasen. *Biochem. Z.* **214**, 1.

[158] 1929. Atmungsferment und Sauerstoffspeicher. *Biochem. Z.* **214**, 4.

[159] 1929. (With F. Kubowitz.) Atmung bei sehr kleinen Sauerstoffdrucken. *Biochem. Z.* **214**, 5.

[160] 1929. (With. F. Kubowitz.) Ist die Atmungshemmung durch Kohlenoxyd vollständig? *Biochem. Z.* **214**, 19.

[161] 1929. (With F. Kubowitz.) Wirkung des Kohlenoxyds auf die Atmung von *Aspergillus oryzae*. *Biochem. Z.* **214**, 24.

[162] 1929. (With E. Negelein and W. Christian.) Über Carbylamin- Hämoglobin und die photochemische Dissoziation seiner Kohlenoxydverbindung. *Biochem. Z.* **214**, 26.

[163] 1929. (With E. Negelein.) Über das Absorptionsspektrum des Atmungsferments. *Biochem. Z.* **214**, 64.

[164] 1929. (With E.Negelein.) Über das Absorptionsspektrum des Atmungsferments der Netzhaut. *Biochem. Z.* **214**, 101.

[165] 1929. (With F. Kubowitz.) Über Atmungsferment im Serum erstickter Tiere. *Biochem. Z.* **214**, 107.

[166] 1929. Über die chemische Konstitution des Atmungsferments *Z. Elektrochem.* **35**, 549.

[167] 1930. (H. A. Krebs.) Manometrische Messung des Kohlensäuregehaltes von Gasgemischen. *Biochem. Z.* **220**, 250.

[168] 1930. (H. A. Krebs.) Manometrische Messung der fermentativen Eiweissspaltung. *Biochem. Z.* **220**, 283.

[169] 1930. (H. A. Krebs.) Versuche über die proteolytische Wirkung des Papains. *Biochem. Z.* **220**, 289.

[170] 1930. (With F. Kubowitz and W. Christian.) Kohlenhydratverbrennung durch Methämoglobin. *Biochem. Z.* **221**, 494.

[171] 1930. (H. L. Alt.) Über die Atmungshemmung durch Blausäure. *Biochem. Z.* **221**, 498.

[172] 1930. (H. Hartmann.) Über das Verhalten von Kohlenoxyd zu Metallverbindungen des Glutathions. *Biochem. Z.* **223**, 489.

[173] 1930. (With E. Negelein and E. Haas.) Spirographishämin. *Biochem. Z.* **227**, 171.

[174] 1930. (With F. Kubowitz.) Über katalytische Wirkung von

Bluthäminen und von Chlorophyll-Häminen. *Biochem. Z.* **227**, 184.

[175] 1930. (With F. KUBOWITZ and W. CHRISTIAN.) Über die katalytische Wirkung von Methylenblau in lebenden Zellen. *Biochem. Z.* **227**, 245.

[176] 1930. Notiz über den Stoffwechsel der Tumoren. *Biochem. Z.* **228**, 257.

[177] 1930. (A. REID.) Oxydation von Leuko-Methylenblau. *Biochem. Z.* **228**, 487.

[178] 1930. (With E. NEGELEIN.) Grünes Haemin aus Blut-Haemin. *Ber. dt. chem. Ges.* **63**, 1816.

[179] 1930. (A. REID.) Über die Oxydation scheinbar autoxydabler Leukobasen durch molekularen Sauerstoff. *Ber. dt. chem. Ges.* **63**, 1920.

[180] 1931. Über Nicht-Hemmung der Zellatmung durch Blausäure. *Biochem. Z.* **231**, 493.

[181] 1931. (With F. KUBOWITZ and W. CHRISTIAN.) Über die Wirkung von Phenylhydrazin und von Phenylhydroxylamin auf die Atmung roter Blutzellen. *Biochem. Z.* **233**, 240.

[182] 1931. Wirkung der Blausäure auf die katalytische Wirkung des Mangans. *Biochem. Z.* **233**, 245.

[183] 1931. (With E. NEGELEIN.) Über die Hauptabsorptionsbanden der MacMunnschen Histohämatine. *Biochem. Z.* **233**, 486.

[184] 1931. (With W. CHRISTIAN.)—Über phäohämin b. *Biochem. Z.* **235**, 240.

[185] 1931. (W. CHRISTIAN.) Aktivierung von Kohlehydrat in roten Blutzellen. *Biochem. Z.* **238**, 131.

[187] 1931. (With E. NEGELEIN.) Photographische Abbildung der Hauptabsorptionsbanden der MacMunnschen Histohämatine. *Biochem. Z.* **238**, 135.

[188] 1931. (A. REID.) Manometrische Messung der sauerstofflosen Atmung. *Biochem. Z.* **242**, 159.

[189] 1931. (With F. KUBOWITZ and W. CHRISTIAN.) Über die Wirkung von Phenylhydrazin und Phenylhydroxylamin auf den Stoffwechsel der roten Blutzellen. *Biochem. Z.* **242**, 170.

[190] 1931. (With W. CHRISTIAN.) Über Aktivierung der Robinsonschen Hexose-Monophosphorsäure in roten Blutzellen und die Gewinnung aktivierender Fermentlösungen. *Biochem. Z.* **242**, 206.

[191] 1931. (With A. REID.) Oxydation von Hämoglobin durch Methylenblau. *Biochem. Z.* **242**, 149.

[192] 1931. Phäophorbid-b-Eisen. *Ber. dt. chem. Ges.* **64**, 682.

[193] 1931. (E. NEGELEIN.) Verbrennung von Kohlenoxyd zu Kohlensäure durch grüne und mischfarbene Hämine. *Biochem. Z.* **243**, 386.

[194] 1931. (With E. NEGELEIN.) Notiz über Spirographishämin. *Biochem. Z.* **244**, 239.

[195] 1932. (With E. NEGELEIN.) Über das Hämin des sauerstoffübertragenden Ferments der Atmung, über einige künstliche Hämoglobine and über Spirographis-Porphyrin. *Biochem. Z.* **244**, 9.

[196] 1932. (With E. NEGELEIN.) Notiz über Spirographishämin. *Biochem. Z.* **244**, 239.

[197] 1932. (E. NEGELEIN.) Kryptohämin. *Biochem. Z.* **248**, 243.

[198] 1932. (E. NEGELEIN.) Über Kryptohämin. *Biochem. Z.* **250**, 577.

[199] 1932. (With W. CHRISTIAN.) Über ein neues Oxydationsferment und sein Absorptionsspektrum. *Biochem. Z.* **254**, 438.

[200] 1932. (F. KUBOWITZ and E. HAAS.) Ausbau der photochemischen Methoden zur Untersuchung des sauerstoffübertragenden Ferments (Anwendung auf Essigbakterien und Hefezellen.) *Biochem. Z.* **255**, 247.

[201] 1932. (H. SCHÜLER.) Über die Oxydation des Hämoglobineisens durch Ferricyankalium und das Gleichgewicht der Reaktion. *Biochem. Z.* **255**, 474.

[202] 1932. (C. V. SMYTHE.) Über die Wirkung von Blausäure auf Methylglyoxal. *Biochem. Z.* **257**, 371.

[203] 1932. (C. V. SMYTHE.) Bildung von Brenztraubensäure aus Methylglyoxal durch katalytische Wirkung der Blausäure. *Ber. dt. chem. Ges.* **65**, 819.

[204] 1932. (C. V. SMYTHE.) Oxydation von Methylglyoxal durch molekularen Sauerstoff bei Gegenwart von Blausäure. *Ber. dt. chem. Ges.* **65**, 1268.

[205] 1932. (With W. CHRISTIAN.) Über das neue Oxydationsferment. *Naturwissenschaften* **20**, 980.

[206] 1932. (With W. CHRISTIAN.) Ein zweites sauerstoffübertragendes Ferment und sein Absorptionsspektrum. *Naturwissenschaften* **20**, 688.

[207] 1932. Das sauerstoffübertragende Ferment der Atmung. *Z. angewandte Chem.* **45**, 1.

[208] 1933. (F. KUBOWITZ and E. HAAS.) Über das Zerstörungsspektrum der Urease. *Biochem. Z.* **257**, 337.

[209] 1933. (With W. CHRISTIAN.) Über das gelbe Oxydationsferment. *Biochem. Z.* **257**, 492.

[210] 1933. (With W. CHRISTIAN.) Über das gelbe Oxydationsferment. *Biochem. Z.* **258**, 496.

[211] 1933. (H. SCHÜLER.) Über das Gleichgewicht zwischen Kohlenoxydhämoglobin und Ferricyankalium. *Biochem. Z.* **259**, 475.

[212] 1933. (C. V. SMYTHE and W. GERISCHER.) Über die Vergärung von Hexosemonophosphorsäure und Glycerinaldehydphosphorsäur. *Biochem. Z.* **260**, 414.

[213] 1933. (With W. CHRISTIAN.) Sauerstoffübertragendes Ferment in Milchsäurebazillen. *Biochem. Z.* **260**, 449.

[214] 1933. (With E. NEGELEIN.) Direkter spektroskopischer Nachweis des sauerstoffübertragenden Ferments in Essigbakterien. *Biochem. Z.* **262**, 237.

[215] 1933. (With W. CHRISTIAN.) Über das gelbe Oxydationsferment. *Biochem. Z.* **263**, 228.

[216] 1933. (H. GAFFRON.) Über den Mechanismus der Sauerstoffaktivierung durch belichtete Farbstoffe. *Biochem. Z.* **264**, 251.

[217] 1933. (W. KEMPNER and F. KUBOWITZ.) Wirkung des Lichtes auf die Kohlenoxydhemmung der Buttersäuregärung. *Biochem. Z.* **265**, 245.

[218] 1933. (With E. NEGELEIN and E. HAAS.) Spektroskopischer Nachweis des sauerstoffübertragenden Ferments neben Cytochrom. *Biochem. Z.* **266**, 1.

[219] 1933. (With W. CHRISTIAN.) Über das gelbe Ferment und seine Wirkungen. *Biochem. Z.* **266**, 377.

[220] 1933. (E. NEGELEIN.) Über die Extraktion eines von Bluthämin verschiedenen Hämins aus dem Herzmuskel. *Biochem. Z.* **266**, 412.

[221] 1933. (E. NEGELEIN and W. GERISCHER.) Direkter spektroskopischer Nachweis des sauerstoffübertragenden Ferments in Azotobakter. *Biochem. Z.* **268**, 1; also *Naturwissenschaften* **21**, 884.

[222] 1934. (W. LÜTTGENS and E. NEGELEIN.) Mikro-Zerewitinoff. *Biochem. Z.* **269**, 177.

[223] 1934. (H. GAFFRON.) Über die Kohlensäureassimilation der roten Schwefelbakterien. I. *Biochem. Z.* **269**, 447.

[224] 1934. (H. THEORELL.) Reindarstellung (Kristallisation) des gelben Atmungsfermentes und die reversible Spaltung desselben. Vorläufige Mitteilung. *Biochem. Z.* **272**, 155.

[225] 1934. (With W. CHRISTIAN.) Co-Fermentproblem. *Biochem. Z.* **274**, 112.

[226] 1934. (F. KUBOWITZ.) Über die Hemmung der Buttersäuregärung durch Kohlenoxyd. *Biochem. Z.* **274**, 285.

[227] 1934. (With E. NEGELEIN.) Cytochrom und sauerstoffübertragendes Ferment. *Naturwissenschaften* **22**, 206.

[228] 1934. (With E. HAAS.) Über eine Absorptionsbande im Gelb in Bäckerhefe. *Naturwissenschaften* **22**, 207.

[229] 1934. Sauerstoffübertragende Fermente. *Naturwissenschaften* **22**, 441.

[230] 1935. (H. THEORELL.) Ein neuer Kataphoreseapparat für Untersuchungszwecke. *Biochem. Z.* **275**, 1.

[231] 1935. (H. THEORELL.) Kataphoretische Studien über das Atmungs-Co-Ferment der roten Blutzellen. *Biochem. Z.* **275**, 11.

[232] 1935. (H. THEORELL.) Bestimmung der Anzahl von sauren Gruppen des Atmungs-Co-Ferments durch Diffusionsmessung. *Biochem. Z.* **275**, 19.

[233] 1935. (H. THEORELL.) Kataphoretische Untersuchungen über die Mischungen von Atmungsfermenten und Substrat. *Biochem. Z.* **275**, 30.

[234] 1935. (H. THEORELL.) Über die Wirkungsgruppe des gelben Ferments. *Biochem. Z.* **275**, 37.

[235] 1935. (With W. CHRISTIAN.) Co-Fermentproblem. *Biochem. Z.* **275**, 464.

[236] 1935. (H. THEORELL.) Reindarstellung der Wirkungsgruppe des gelben Fermentes. *Biochem. Z.* **275**, 344.

[237] 1935. (H. THEORELL.) Über Hemmung der Reaktionsgeschwindigkeit durch Phosphat in Warburg's und Christian's System. *Biochem. Z.* **275**, 416.

[238] 1935. (H. THEORELL.) Das gelbe Oxydationsferment. *Biochem. Z.* **278**, 263.

[239] 1935. (H. THEORELL.) Ein Kataphorese-apparat für präparative Zwecke. *Biochem. Z.* **278**, 291.

[240] 1935. (With W. CHRISTIAN and A. GRIESE.) Die Wirkungsgruppe des Co-Ferments aus roten Blutzellen. *Biochem. Z.* **279**, 143.

[241] 1935. (H. THEORELL.) Quantitative Bestrahlungsversuche an gelbem Ferment, Flavinphosphorsäure und Lactoflavin. *Biochem. Z.* **279**, 186.

[242] 1935. (W. LÜTTGENS and W. CHRISTIAN.) Ultrakohlenstoffbestimmung. *Biochem. Z.* **281**, 310.

[243] 1935. (With W. CHRISTIAN and A. GRIESE.) Wasserstoff-übertragendes Co-Ferment, seine Zusammensetzung und Wirkungsweise, *Biochem. Z.* **282**, 157.

[244] 1935. (E. NEGELEIN and E. HAAS.) Über die Wirkungsweise des Zwischenferments. *Biochem. Z.* **282**, 206.

[245] 1935. (With W. CHRISTIAN.) Zerstörung des wasserstoff-übertragenden Co-Ferments durch ultraviolettes Licht. *Biochem. Z.* **282**, 221.

[246] 1935. (E. HAAS.) Über das Absorptionsspektrum des Wassers im Ultraviolett. *Biochem. Z.* **282**, 224.

[247] 1935. (F. KUBOWITZ.) Kohlenoxyd-Ferroglutathion. *Biochem. Z.* **282**, 277.

[248] 1936. (With W. CHRISTIAN.) Pyridin, der wasserstoff-übertragende Bestandteil von Gärungsfermenten. *Helv. chim. Acta* **19**, E, 79.

[249] 1935. (E. NEGELEIN and W. GERISCHER.) Verbesserte Methode zur Gewinnung des Zwischenferments aus Hefe. *Biochem. Z.* **284**, 289.

[250] 1936. (With W. CHRISTIAN.) Gärungs-Co-Ferment. *Biochem. Z.* **285**, 156.

[251] 1936. (With P. KARRER.) Jodmethylat des Nicotinsäure-amids. *Biochem. Z.* **285**, 297.

[252] 1936. (E. HAAS.) Mikrohydrierung mit Hydrosulfit. *Biochem. Z.* **285**, 368.

[253] 1936. (With W. CHRISTIAN.) Optischer Nachweis der Hydrierung und Dehydrierung des Pyridins im Gärungs-Co-Ferment. *Biochem. Z.* **286**, 81.

[254] 1936. (With W. CHRISTIAN.) Pyridin als Wirkungsgruppe dehydrierender Fermente. *Biochem. Z.* **286**, 142.

[255] 1936. (With W. CHRISTIAN.) Pyridin, der wasserstoff-übertragende Bestandteil von Gärungsfermenten (Pyridin-Nucleotide.) *Biochem. Z.* **287**, 291.

[256] 1936. (E. NEGELEIN.) Methode zur Gewinnung des A-Proteins der Gärungsfermente. *Biochem. Z.* **287**, 329.

[257] 1936. (With W. CHRISTIAN.) Verbrennung von Robison-Ester durch Triphospho-Pyridin-Nucleotid. *Biochem. Z.* **287**, 440.

[258] 1936. (E. HAAS.) Verhalten der Absorptionsspektren der Dihydro-pyridinverbindungen. *Biochem. Z.* **288**, 123.

[259] 1936. Enzymes as oxygen carriers. *Inst. Intern. Chim. Solvay. 5th Conference*, p. 303.

[260] 1936. (With W. CHRISTIAN.) Über Nikotinsäureamid und Luminoflavin. *Ber. dt. chem. Ges.* **69B**, 228.

[261] 1937. (E. NEGELEIN and H.-J. WULFF.) Kristallisation des Proteins der Acetaldehydreduktase. *Biochem. Z.* **289**, 436.

[262] 1937. (E. HAAS.) Wirkungsweise des Proteins des gelben Ferments. *Biochem. Z.* **290**, 291.

[263] 1937. (E. NEGELEIN and H.-J. WULFF.) Dissoziationskonstanten und Reaktionsfähigkeit der Acetaldehydreduktase. *Biochem. Z.* **290**, 445.

[264] 1937. (E. HAAS.) Manometrische Mikrotitration mit Ferricyanid. *Biochem. Z.* **291**, 79.

[265] 1937. (C. S. FRENCH.) The quantum yield of hydrogen and carbon dioxide assimilation in purple bacteria. *J. gen. Physiol.* **20**, 711.

[266] 1937. (C. S. FRENCH.) The rate of CO_2 assimilation by purple bateria at various wave lengths of light. *J. gen. Physiol.* **21**, 71.

[267] 1937. (F. KUBOWITZ.) Über die chemische Zusammensetzung der Kartoffeloxydase. *Biochem. Z.* **292**, 221.

[268] 1937. (With W. CHRISTIAN.) Abbau von Robison-Ester durch Triphospho-Pyridin Nucleotid. *Biochem. Z.* **292**, 287.

[269] 1937. (F. KUBOWITZ.) Schwermetallproteid und Pyridinproteid, die Komponenten Blausäure—und Kohlenoxyd-empfindlicher Alkoholdehydrasen. *Biochem. Z.* **293**, 308.

[270] 1937. (E. NEGELEIN and H.-J. WULFF.) Diphosphopyridinproteid. Alkohol, Acetaldehyd. *Biochem. Z.* **293**, 351.

[271] 1938. (With W. CHRISTIAN.) Co-Ferment der d-Aminosäure-Deaminase. *Biochem. Z.* **295**, 261.

[272] 1938. (E. G. BALL.) Über die Oxydation und Reduktion der drei Cytochrom-Komponenten. *Biochem. Z.* **295**, 262.

[273] 1938. (With W. CHRISTIAN.) Co-Ferment der *d*-Alanin-Oxydase. *Biochem. Z.* **296**, 294.

[274] 1938. (F. KUBOWITZ.) Re-Synthese der Phenoloxydase aus Protein und Kupfer. *Biochem. Z.* **296**, 443.

[275] 1938. (With W. CHRISTIAN and A. GRIESE.) Alloxazin-Adenin-Dinukleotid aus Hefe. *Biochem. Z.* **297**, 417.

[276] 1938. (With W. CHRISTIAN.) Isolierung der prosthetischen

Gruppe der d-Aminosäureoxydase. *Biochem. Z.* **298**, 150.

[277] 1938. (With W. CHRISTIAN.) Bemerkung über gelbe Fermente. *Biochem. Z.* **298**, 368.

[278] 1938. (E. HAAS.) Isolierung eines neuen gelben Ferments. *Biochem. Z.* **298**, 378.

[279] 1938. (F. KUBOWITZ.) Spaltung und Resynthese der Polyphenoloxydase und des Hämocyanins. *Biochem. Z.* **299**, 32.

[280] 1938. (E. G. BALL.) Xanthine oxidase: an alloxazine proteid. *Science N.Y.* **88**, 131.

[281] 1938. (J. N. DAVIDSON,) Purified uricase. *Nature, Lond.* **141**, 790.

[282] 1938. (With W. CHRISTIAN.) Koferment der d-Alanin-Dehydrase. *Naturwissenschaften* **26**, 201.

[283] 1938. Chemische Konstitution von Fermenten. *Ergebn. Enzymforsch.* **7**, 210.

[284] 1939. (E. NEGELEIN and H. BRÖMEL.) Protein der *d*-Aminosäureoxydase. *Biochem. Z.* **300**, 225.

[285] 1939. (E. NEGELEIN and H. BRÖMEL.) Isolierung eines reversiblen Zwischenprodukts der Gärung. *Biochem. Z.* **301**, 135.

[286] 1939. (With W. CHRISTIAN.) Proteinteil des kohlenhydratoxydierenden Ferments der Gärung. *Biochem. Z.* **301**, 221.

[287] 1939. (With W. CHRISTIAN.) Isolierung und Kristallisation des Proteins des oxydierenden Gärungsferments. *Biochem. Z.* **303**, 40.

[288] 1939. (E. NEGELEIN and H. BRÖMEL.) R-Diphosphoglycerinsäure, ihre Isolierung und Eigenschaften. *Biochem. Z.* **303**, 132.

[289] 1939. (E. NEGELEIN and H. BRÖMEL.) Über die Entstehung von Glycerin bei der Gärung. *Biochem. Z.* **303**, 231.

[290] 1939. (E. G. BALL.) Xanthine oxidase: purification and properties. *J. biol. Chem.* **128**, 51.

[291] 1941. (F. KUBOWITZ and W. LÜTTGENS.) Zusammensetzung, Spaltung und Resynthese der Carboxylase. *Biochem. Z.* **307**, 170.

[292] 1941. (With W. CHRISTIAN.) Isolierung und Kristallisation des Gärungsferments Enolase. *Biochem. Z.* **310**, 384.

[293] 1941. (With W. CHRISTIAN.) Isolierung und Kristallisation des Gärungsferments Enolase. *Naturwissenschaften* **29**, 589.

[294] 1941. (With W. CHRISTIAN.) Chemischer Mechanismus der Fluorid-Hemmung der Gärung. *Naturwissenschaften* **29**, 590.

[295] 1941. (T. BÜCHER and E. NEGELEIN.) Versuche zum Quanten-Problem der Kohlensäure-Assimilation. *Naturwissenschaften* **29**, 591.

[296] 1941. (E. NEGELEIN and T. BÜCHER.) Photochemische Spaltung von Kohlenoxyd-Myoglobin. *Naturwissenschaften* **29**, 672.

[297] 1942. (With W. CHRISTIAN.) Isolierung und Kristallisation des Gärungsferments Enolase. *Biochem. Z.* **310**, 384.

[298] 1942. (T. BÜCHER and E. NEGELEIN.) Photochemische Ausbeute bei der Spaltung des Kohlenoxyd-Hämoglobins. *Biochem. Z.* **311**, 163.

[299] 1942. (With W. CHRISTIAN.) Wirkungsgruppe des Gärungsferments Zymohexase. *Biochem. Z.* **311**, 209.

[300] 1942. (F. KUBOWITZ and P. OTT.) Isolierung und Kristallisation eines Gärungsferments aus Tumoren. *Biochem. Z.* **314**, 94.

[301] 1942. (With W. CHRISTIAN.) Zymohexase im Blutplasma von Tumortieren. *Naturwissenschaften* **30**, 731.

[302] 1942. (With W. CHRISTIAN.) Isolierung und Kristallisation des Gärungsferments Zymohexase. *Naturwissenschaften* **30**, 731.

[303] 1942. (F. KUBOWITZ and P. OTT.) Tumorferment und Muskelferment. *Naturwissenschaften* **30**, 732.

[304] 1942. (T. BÜCHER.) Isolierung und Kristallisation eines phosphatübertragenden Gärungsferments. *Naturwissenschaften* **30**, 756.

[305] 1943. (With W. CHRISTIAN.) Isolierung und Kristallisation des Gärungsferments Zymohexase. *Biochem. Z.* **314**, 149.

[306] 1943. (With W. CHRISTIAN.) Gärungsfermente in Blutserum von Tumor-Ratten. *Biochem. Z.* **314**, 399.

[307] 1943. (T. BÜCHER.) Optische Molekulargewichtsbestimmung an kristallisierten Ferment- Proteinen. *Angew. Chem.* **56**, 328.

[308] 1944. (F. KUBOWITZ and P. OTT.) Isolierung von Gärungsfermenten aus menschlichen Muskeln. *Biochem. Z.* **317**, 193.

[309] 1944. (With W. LÜTTGENS.) Experiment zur Assimilation der Kohlensäure. *Naturwissenschaften* **32**, 161.

[310] 1944. (With W. LÜTTGENS.) Weitere Experimente zur

Kohlensäureassimilation. *Naturwissenschaften* **32**, 301.

[311] 1944. (T. BÜCHER and J. KASPERS.) Lichtfilter für 280 mμ. *Naturwissenschaften* **33**, 93.

[312] 1944. (T. BÜCHER and J. KASPERS.) Photochemische Spaltung des Kohlenoxyd-Myoglobins durch ultraviolettes Licht. *Naturwissenschaften* **33**, 93.

[313] 1946. Molekulargewicht des sauerstoffübertragenden Ferments. *Naturwissenschaften* **33**, 94.

[314] 1946. Über den Quantenbedarf der Kohlensäureassimilation. *Naturwissenschaften* **33**, 122.

[315] 1946. (With W. LÜTTGENS.) Photochemische Reduktion des Chinons in grünen Zellen und Granula. *Biochimia* **11**, 303.

[316] 1947. (T. BÜCHER and J. KASPERS.) Photochemische Spaltung des Kohlenoxydmyoglobins durch ultraviolette Strahlung (Wirksamkeit der durch die Proteinkomponente des Pigments absorbierten Quanten). *Biochim. biophys. Acta* **1**, 21.

[317] 1947. (T. BÜCHER.) Über ein phosphatübertragendes Gärungsferment. *Biochim. biophys. Acta* **1**, 292.

[318] 1947. (T. BÜCHER.) Über das Molekulargewicht der Enolase. *Biochim. biophys. Acta* **1**, 467.

[319] 1947. (T. BÜCHER.) Zum Gültigkeitsbereich der Rayleighschen Gleichung (Molekulargewichte aus der Lichtzerstreuung von Linearkolloiden). *Biochim. biophys. Acta* **1**, 477.

[320] 1947. Ideen zur Fermentchemie der Tumoren. *Abhandlg. Dtsch. Akad. Wiss. Berlin* **3**, 3.

[321] 1948. Assimilatory quotient and photochemical yield. *Am. J. Bot.* **35**, 194.

[322] 1949. (With D. BURK, V. SCHOCKEN, S. HENDRICKS, and M. KORZENOVSKY.) The maximum efficiency of photosynthesis. *Science, N.Y.* **110**, 225.

[323] 1949. (With V. SCHOCKEN.) A manometric actinometer for the visible spectrum. *Arch. Biochem.* **21**, 363.

[324] 1949. (With D. BURK, V. SCHOCKEN, M. KORZENOVSKY, and S. B. HENDRICKS.) Does light inhibit the respiration of green cells? *Arch. Biochem.* **23**, 330.

[325] 1950. (With D. BURK.) The maximum efficiency of photosynthesis. *Arch. Biochem.* **25**, 410.

[326] 1950. (With D. BURK.) 1-Quanten-Mechanismus und Energie-Kreisprozess bei der Photosynthese. *Naturwissenschaften* **37**, 560.

[327] 1950. (With D. BURK, S. B. HENDRICKS, and V. SCHOCKEN.) The quantum efficiency of photosynthesis. *Biochim. biophys. Acta* **4**, 335.

[328] 1951. 1-Quanten-Mechanismus der Photosynthese. *Z. Elektrochem. angew. phys. Chem.* **55**, 447.

[329] 1951. (With D. BURK.) Ein-Quanten-Reaktion und Kreisprozess der Energie bei der Photosynthese. *Z. Naturforsch.* **6b**, 12.

[330] 1951. (With H. GELEICK.) Über den Gewinn im Kreisprozess der Photosynthese. *Z. Naturforsch.* **6b**, 134.

[331] 1951. (With H. GELEICK and K. BRIESE.) Über die Aufspaltung der Photosynthese in Lichtreaktion und Rückreaktion. *Z. Naturforsch.* **6b**, 417.

[332] 1951. (With H. GELEICK and K. BRIESE.) Weitere Steigerung des Energiegewinns im Kreisprozess der Photosynthese. *Z. Naturforsch.* **6b**, 285.

[333] 1951. (With H. GELEICK and K. BRIESE.) Über die Aufspaltung der Photosynthese in Lichtreaktion und Rückreaktion. *Z. Naturforsch.* **6b**, 417.

[334] 1951. (With H.-S. GEWITZ.) Cytohämin aus Herzmuskel. *Hoppe-Seyler's Z. physiol. Chem.* **288**, 1.

[335] 1952. Energetik der Photosynthese. *Naturwissenschaften* **39**, 337.

[336] 1952. Der Mechanismus des Photolyse-Photosynthese-Systems der grünen Pflanzen. *Naturwissenschaften* **39**, 185.

[337] 1952. (With H. GELEICK and K. BRIESE.) Über die Messung der Photosynthese in Carbonat-Bicarbonat-Gemischen. *Z. Naturforsch.* **7b**, 141.

[338] 1952. (With E. HIEPLER.) Versuche mit Ascites-Tumorzellen. *Z. Naturforsch.* **7b**, 193.

[339] 1952. Über den Einfluss der Wellenlänge auf die Chinonreduktion in grünen Grana. *Z. Naturforsch.* **7b**, 443.

[340] 1952. (With D. BURK and A. L. SCHADE.) Extensions of photosynthetic experimentation. In *Symposia of the Society for Experimental Biology V. Carbon dioxide fixation and photosynthesis*, p. 306. Cambridge University Press.

[341] 1952. (H. TIEDEMANN.) Über den Stoffwechsel des Ascites-Tumors der Maus. *Z. ges. exptl. Med.* **119**, 272.

[342] 1953. (With K. DAMASCHKE, F. TÖDT, and D. BURK.) An electrochemical demonstration of the energy cycle and maximum quantum yield in photosynthesis. *Biochim. biophys. Acta* **12**, 347.

[343] 1953. (With G. KRIPPAHL, W. SCHRÖDER, and W. BUCHHOLZ.) Messung des Quantenbedarfs der Photosynthese für sehr dünne Zellsuspensionen. *Biochim. biophys. Acta* **12**, 356.

[344] 1953. (With G. KRIPPAHL, W. BUCHHOLZ, and W. SCHRÖDER.) Weiterentwicklung der Methoden zur Messung der Photosynthese. *Z. Naturforsch.* **8b**, 675.

[345] 1953. Über die Wirkungsgruppen der oxydierenden und reduzierenden Fermente. *Naturwissenschaften* **40**, 493.

[346] 1953. Über das Verhalten von Ascites-Tumorzellen zu Sauerstoff von höheren Drücken. *Arch. Geschwulstorsch* **6**, 7.

[347] 1953. (With H.-S. GEWITZ.) Cyto-Deutero-Porphyrin. *Hoppe-Seyler's Z. physiol. Chem.* **292**, 174.

[348] 1954. (With K. GAWEHN and G. LANGE.) Über das Verhalten von Ascites-Krebszellen zu Zymohexase. *Z. Naturforsch.* **9b**, 109.

[349] 1954. (With G. KRIPPAHL, W. SCHRÖDER, W. BUCHHOLZ, and E. THEEL.) Über die Wirkung sehr schwachen blaugrünen Lichts auf den Quantenbedarf der Photosynthese. *Z. Naturforsch.* **9b**, 164.

[350] 1954. (With G. KRIPPAHL.) Messung der Lichtabsorption in *Chlorella* mit der Ulbrichtschen Kugel. *Z. Naturforsch.* **9b**, 181.

[351] 1954. (With K. GAWEHN.) Isolierung der Hefezymohexase und ihre Kristallisation als Quecksilbersalz. *Z. Naturforsch.* **9b**, 206.

[352] 1954. Über die Berücksichtigung der Retention der Kohlensäure bei Messungen der Photosynthese in Kulturlösungen. *Z. Naturforsch.* **9b**, 302.

[353] 1954. (With H. KLOTZSCH and K. GAWEHN.) Über die Oxydationsreaktion der Gärung. *Z. Naturforsch.* **9b**, 391.

[354] 1954. Über Phosphorylierung durch Dehydrierung. Supplement to *Z. Naturforsch.* **9b**, 391.

[355] 1954. (With G. KRIPPAHL and W. SCHRÖDER.) Katalytische Wirkung des blaugrünen Lichts auf den Energieumsatz bei der Photosynthese. *Z. Naturforsch.* **9b**, 667.

[356] 1954. (With G. KRIPPAHL, W. SCHRÖDER, and W. BUCHHOLZ.) Sauerstoff-Kapazität der *Chlorella*. *Z. Naturforsch.* **9b**, 769.

[357] 1954. (With G. KRIPPAHL.) Über Photosynthese-Fermente. *Angew. Chem.* **66**, 493.

[358] 1954. Krebsforschung. *Naturwissenschaften* **41**, 485.

[359] 1955. (G. KRIPPAHL.) Carotinoid als Wirkungsgruppe eines

Photosynthese-Ferments. *Ann. Acad. Scient. Fennicae* **60**, 69.

[360] 1955. (With G. KRIPPAHL.) Photochemische Wasserzersetzung durch lebende *Chlorella*. *Z. Naturforsch.* **10b**, 301.

[361] 1955. (With G. KRIPPAHL and W. BUCHHOLZ.) Wirkung von Vanadium auf die Photosynthese. *Z. Naturforsch.* **10b**, 422.

[362] 1955. (With H.-S. GEWITZ and W. VÖLKER.) Über Gewinnung und Abbau des Cytohämins. *Z. Naturforsch.* **10b**, 541.

[363] 1955. (With G. KRIPPAHL and W. SCHRÖDER.) Wirkungsspektrum eines Photosynthese-Ferments. *Z. Naturforsch.* **10b**, 631.

[364] 1955. (With W. SCHRÖDER.) Versuche über die Sauerstoff-Kapazität der *Chlorella*. *Z. Naturforsch.* **10b**, 639.

[365] 1955. (H.-S. GEWITZ and W. VÖLKER.) Über die Synthese einiger Deuteroporphyrine. *Hoppe-Seyler's Z. physiol. Chem.* **302**, 119.

[366] 1955. (With G. KRIPPAHL.) Activity spectrum of a photosynthetic enzyme. *Pubbl. Staz. Zool. Napoli.* **27**, 1.

[367] 1955. Über die Messung des Energieumsatzes bei der Photosynthese mit dem Grossflächen-Bolometer. *Biochim. biophys. Acta* **18**, 163.

[368] 1955. Über die Entstehung der Krebszellen. *Naturwissenschaften* **42**, 401.

[369] 1955. Photodissoziation und induzierte Atmung, die Fundamental-Reaktionen der Photosynthese. *Naturwissenschaften* **42**, 449.

[370] 1955. I. Über die Entstehung der Krebszellen. II. Experimente über Photosynthese. *Mitteilungen aus der Max-Planck-Gesellschaft* **4**, 166.

[371] 1955. Über die Entstehung der Krebszellen. In *Krebsforschung und Krebsbekämpfung* (*Sonderbände zur Strahlentherapie*, **34**), p. 3.

[372] 1956. (With G. KRIPPAHL.) Über die funktionelle Kohlensäure der *Chlorella*. *Z. Naturforsch.* **11b**, 718.

[373] 1956. (With G. KRIPPAHL.) Über die CO_2-Kapazität der *Chlorella* und den chemischen Mechanismus der CO_2—Assimilation. *Z. Naturforsch.* **11b**, 52.

[374] 1956. (With G. KRIPPAHL.) Über die funktionelle Carboxylgruppe des Chlorophylls. *Z. Naturforsch.* **11b**, 179.

[375] 1956. (With W. SCHRÖDER and H.-W. GATTUNG.) Züchtung

der *Chlorella* mit fluktuierender Lichtintensität. *Z. Naturforsch.* **11b,** 654.

[376] 1956. (With K. GAWEHN and A.-W. GEISSLER.) Stoffwechsel von embryonalen Zellen und von Krebszellen. *Z. Naturforsch.* **11b,** 657.

[377] 1956. (With G. KRIPPAHL.) Über die funktionelle Kohlensäure der *Chlorella*. *Z. Naturforsch.* **11b,** 718.

[378] 1956. On the origin of cancer cells. *Science, N.Y.* **123,** 309.

[379] 1956. On respiratory impairment in cancer cells. *Science, N.Y.* **124,** 269.

[380] 1956. (With G. KRIPPAHL and W. SCHRÖDER.) Über den chemischen Mechanismus der Kohlensäureassimilation. *Naturwissenschaften* **43,** 237.

[381] 1956. (G. KRIPPAHL and W. SCHRÖDER.) Über den chemischen Mechanismus der Kohlensäure-Assimilation. *Angew. Chem.* **68,** 418.

[382] 1956. Über die Entstehung der Krebszellen. *Oncologia* **9** (2), 75.

[383] 1956. Krebsforschung. From '*Aus der Deutschen Forschung der letzen Dezennien*'. Dr. Ernst Telschow zum 65. Geburtstag gewidmet (ed. B. Rajewsky and G. Schreiber). Georg Thieme Verlag, Stuttgart.

[384] 1956. (D. BURK and A. L. SCHADE.) On respiratory impairment in cancer cells. *Science, N.Y.* **124,** 270.

[385] 1956. Chemischer Mechanismus der CO_2—Assimilation und die Theorie von Willstätter-Stoll. *Festschrift Arthur Stoll*, p. 705. Sandoz AG, Basel.

[387] 1957. Stoffwechsel von embryonalen Zellen in Gewebekulturen. *Biochem. biophys. Acta* **25,** 429.

[388] 1957. (With K. GAWEHN and A.-W. GEISSLER.) Manometrie der Körperzellen unter physiologischen Bedingungen. *Z. Naturforsch.* **12b,** 115.

[389] 1957. (With K. GAWEHN and A.-W. GEISSLER.) Stöchiometrische Versuche mit dem oxydierenden Gärungsferment. *Z. Naturforsch.* **12b,** 47.

[390] 1957. (H. KLOTZSCH and G. KRIPPAHL.) Über die Funktion der Glutaminsäure in Chlorella. *Z. Naturforsch.* **12b,** 266.

[391] 1957. (With K. GAWEHN and A.-W. GEISSLER.) Über die Wirkung von Wasserstoffperoxyd auf Krebszellen und auf embryonale Zellen. *Z. Naturforsch.* **12b,** 393.

[392] 1957. (With H. KLOTZSCH and G. KRIPPAHL.) Über das Verhalten einiger Aminosäuren in Chlorella bei

Zusatz von markierter Kohlensäure. *Z. Naturforsch.* **12b**, 481.

[393] 1957. (With H. KLOTZSCH and G. KRIPPAHL.) Glutaminsäure in *Chlorella*. *Z. Naturforsch.* **12b**, 622.

[394] 1957. (With W. SCHRÖDER.) Quantenbedarf der Photosynthese. *Z. Naturforsch.* **12b**, 716.

[395] 1957. (With H.-S. GEWITZ and W. VÖLKER.) D(—) Milchsäure in *Chlorella*. *Z. Naturforsch.* **12b**, 722.

[396] 1957. (With W. SCHRÖDER, G. KRIPPAHL, and H. KLOTSZCH.) Photosynthese. *Angew. Chem.* **69**, 627.

[397] 1957. (D. BURK.) Über die Begründung einer Chemotherapie des Krebses auf der Grundlage einer primären Hemmung der Glucosephosphorylierung. *Klin. Wschr.* **35**, 1102.

[398] 1957. (With H. KLOTZSCH and G. KRIPPAHL.) Glutaminsäure-Decarboxylase in *Chlorella*. *Naturwissenschaften* **44**, 235.

[399] 1957. (With G. KRIPPAHL.) Ausbau der manometrischen Methoden. *Justus Liebigs Annalen Chem.* **604**, 94.

[400] 1957. Paul Ehrlich 1854–1915. In *Die Grossen Deutschen*, Bd. 4, p. 186. Propyläen Verlag, Berlin.

[401] 1957. Atmungshemmung und Karzinogenese. *Experientia.* **13**, 125.

[402] 1958. (With K. GAWEHN and A.-W. GEISSLER.) Über die Entstehung des Krebsstoffwechsels in der Gewebekultur. *Z. Naturforsch.* **13b**, 61.

[403] 1958. (With G. KRIPPAHL.) Beweis der Notwendigkeit der Glutaminsäure für die Photosynthese. *Z. Naturforsch.* **13b**, 63.

[404] 1958. (With G. KRIPPAHL.) Sauerstoff-Halbwertdrucke der Photosynthese und Atmung. *Z. Naturforsch.* **13b**, 66.

[405] 1958. (With G. KRIPPAHL.) Weiterentwicklung der manometrischen Methoden. *Z. Naturforsch.* **13b**, 434.

[406] 1958. (With G. KRIPPAHL, H.-S. GEWITZ, and W. VÖLKER.) Carotinoid-Oxygenase in *Chlorella*. *Z. Naturforsch.* **13b**, 437.

[407] 1958. (With G. KRIPPAHL.) Hill-Reaktionen. *Z. Naturforsch.* **13b**, 509.

[408] 1958. (With K. GAWEHN and A.-W. GEISSLER.) Stoffwechsel der weissen Blutkörperchen. *Z. Naturforsch.* **13b**, 515.

[409] 1958. (With K. GAWEHN and A.-W. GEISSLER.) Katalasegehalt, Gärung und Atmung in der Zellkultur nach Dulbecco- Vogt. *Z. Naturforsch.* **13b**, 588.

[410] 1958. (With W. Schröder, H.-S. Gewitz, and W. Völker.) Manometrisches Röntgenstrahlen-Aktinometer und über die Wirkung der Röntgenstrahlen auf die Gärung von Krebszellen. *Z. Naturforsch.* **13b**, 591.

[411] 1958. (With G. Krippahl, H.-S. Gewitz, and W. Völker.) Über den chemischen Mechanismus der Photosynthese. *Z. Naturforsch.* **14b**, 712.

[412] 1958. (With G. Krippahl and W. Schröder.) Über den chemischen Mechanismus der Kohlensäureassimilation. *Naturwissenschaften* **43**, 237.

[413] 1958. (With W. Schröder, H. Gewitz, and W. Völker.) Über die selektive Wirkung der Röntgenstrahlen auf Krebszellen. *Naturwissenschaften* **45**, 192.

[414] 1958. (With W. Schröder, G. Krippahl, and H. Klotzsch.) Photosynthesis. *Angew. Chem.* **669**, 627.

[415] 1958. (With K. Gawehn, A.-W. Geissler, W. Schröder, H. Gewitz, and W. Völker.) Partielle Anaerobiose und Strahlenempfindlichkeit der Krebszellen. *Arch. Biochem. Biophys.* **78**, 573.

[416] 1958. Photosynthesis. *Science, N.Y.* **128**, 68.

[417] 1959. (With K. Gawehn, A.-W. Geissler, W. Schröder, H. Gewitz, and W. Völker.) Partielle Anaerobiose der Krebszellen und Wirkung der Röntgenstrahlen auf Krebszellen. *Naturwissenschaften* **46**, 25.

[418] 1959. (With G. Krippahl, H. Gewitz, and W. Völker.) Über den chemischen Mechanismus der Photosynthese. *Angew. Chem.* **71**, 633.

[419] 1959. (With G. Krippahl.) Weiterentwicklung der manometrischen Methoden. *Z. Naturforsch.* **14b**, 561.

[420] 1959. (With D. Kayser.) Wirkung von Kohlenoxyd auf Atmung und Photosynthese in grünen Keimblättern. *Z. Naturforsch.* **14b**, 563.

[421] 1959. (With G. Krippahl, H.-S. Gewitz, and W. Völker.) Über den chemischen Mechanismus der Photosynthese. *Z. Naturforsch.* **14b**, 712.

[422] 1959. (With G. Krippahl.) Further development of manometric methods. *J. natn. Cancer Inst.* **24**, 51.

[423] 1960. (With W. Schröder and H.-W. Gattung.) Über die Wirkung von Röntgenstrahlen auf Hämoglobin (Mit einer Bemerkung über die Strahlenempfindlichkeit von Gewebeschnitten.) *Z. Naturforsch.* **15b**, 163.

[424] 1960. (With G. Krippahl.) Neubestimmung des Quanten-

bedarfs der Photosynthese mit der Kompensationsmethode. *Z. Naturforsch.* **15b**, 190.

[425] 1960. (With G. KRIPPAHL.) Glykolsäurebildung in *Chlorella*. *Z. Naturforsch.* **15b**, 197.

[426] 1960. (With G. KRIPPAHL.) Weiterentwicklung der manometrischen Methoden. (Carbonatgemische.) *Z. Naturforsch.* **15b**, 364.

[427] 1960. (With G. KRIPPAHL.) Notwendigkeit der Kohlensäure für die Chinon- und Ferricyanid-Reaktionen in grünen Grana. *Z. Naturforsch.* **15b**, 367.

[428] 1960. (With G. KRIPPAHL and H.-W. GATTUNG.) 1-Gefäss-Methode zur Messung des Quantenbedarfs der Photosynthese. *Z. Naturforsch.* **15b**, 370.

[429] 1960. (With K. GAWEHN and A.-W. GEISSLER.) Umwandlung des embryonalen Stoffwechsels in Krebsstoffwechsel. *Z. Naturforsch.* **15b**, 378.

[430] 1960. (H.-S. GEWITZ and W. VÖLKER.) Weiterentwicklung der manometrischen Methoden (Herstellung definierter Blausäure-Konzentrationen). *Z. Naturforsch.* **15b**, 625.

[431] 1960. (With G. KRIPPAHL.) Weiterentwicklung der manometrischen Methoden (Wechsel aerober und anaerober Bedingungen.) *Z. Naturforsch.* **15b**, 786.

[432] 1960. (With G. KRIPPAHL.) Über den Photolyten der Photosynthese. *Z. Naturforsch.* **15b**, 788.

[433] 1960. (With K. GAWEHN and A.-W. GEISSLER.) Weiterentwicklung der zellphysiologischen Methoden (Verbindung von Manometrie und optischer Milchsäurebestimmung). *Hoppe-Seyler's Z. Physiol. Chem.* **320**, 277.

[434] 1960. (With K. GAWEHN, A.-W. GEISSLER, and S. LORENZ.) Über die Umwandlung des embryonalen Stoffwechsels in Krebsstoffwechsel. *Hoppe-Seyler's Z. physiol. Chem.* **321**, 252.

[435] 1961. Geschichte des Max-Planck-Instituts für Zellphysiologie. *Jahrbuch d. Max-Planck-Gesellschaft* II, 816.

[436] 1961. (With A.-W. GEISSLER and S. LORENZ.) CO_2-Drucke über Bicarbonat-Carbonatgemischen. *Z. Naturforsch.* **16b**, 283.

[437] 1961. (H.-S. GEWITZ and W. VÖLKER.) 3-Hydroxytyramin, ein biologischer Oxydationskatalysator der Photosynthese. *Z. Naturforsch.* **16b**, 559.

[438] 1961. Über die fakultative Anaerobiose der Krebszellen und

ihre Anwendung auf die Chemotherapie. *Münch. Med. Wschr.* **103,** 2504.

[439] 1962. (With G. KRIPPAHL, A.-W. GEISSLER, and S. LORENZ.) Weiterentwicklung der manometrischen Methoden (Sauerstoffabsorption durch Chromchlorür). *Z. Naturforsch.* **17b,** 17.

[440] 1962. (With G. KRIPPAHL.) Messung von Autoxydationen. *Z. Naturforsch.* **17b,** 281.

[442] 1962. (With G. KRIPPAHL, A.-W. GEISSLER, and S. LORENZ.) Sauerstoffabsorption durch Chromchlorür. *Z. Naturforsch.* **17b,** 281.

[443] 1962. (With G. KRIPPAHL.) Züchtung von *Chlorella* mit der Xenon-Hochdrucklampe. *Z. Naturforsch.* **17b,** 631.

[444] 1962. (With G. KRIPPAHL.) Weiterentwicklung der manometrischen 1-Gefässmethoden. *Z. Naturforsch.* **17b,** 631.

[445] 1962. (D. KAYSER.) Über die Messung des Sauerstoffdruckes, bei dem Sporen von *Clostridium butyricum* auskeimen. *Z. Naturforsch.* **17b,** 658.

[446] 1962. (With A.-W. GEISSLER and S. LORENZ.) Atmung und Wachstum von Krebszellen bei niedrigen Sauerstoffdrucken. *Z. Naturforsch.* **17b,** 758.

[447] 1962. (With K. GAWEHN, A.-W. GEISSLER and S. LORENZ.) Über den Einfluss des Sauerstoffdrucks auf die Umwandlung des embryonalen Stoffwechsels in Krebsstoffwechsel. *Z. Naturforsch.* **17b,** 772.

[448] 1962. On phosphorylation in light. *J. gen. Physiol.* **45,** (Pt. 2), 17.

[449] 1962. In La Photosynthèse. *Colloques Internationaux du Centre National de la Recherche Scientifique,* **119,** 221.

[450] 1963. (B. VENNESLAND, H. W. GATTUNG, and E. BIRKICHT.) Fluorometric measurement of the photoreduction of flavin by illuminated chloroplasts. *Biochim. biophys. Acta* **66,** 285.

[451] 1963. (D. KAYSER.) Über die Hemmung des Wachstums von Bacterium lactis aerogenes durch Röntgenstrahlen. *Z. Naturforsch.* **18b,** 171.

[452] 1963. (With G. KRIPPAHL.) Optische Bestimmung von Wasserstoffperoxyd mit Vanadinsäure. *Z. Naturforsch.* **18b,** 340.

[453] 1963. (H.-S. GEWITZ and W. VÖLKER.) Hemmung der Photosynthese durch Kohlenoxyd und Aufhebung der Kohlenoxydhemmung durch Licht. *Z. Naturforsch.* **18b,** 649.

[454] 1963. (With K. GAWEHN, A.-W. GEISSLER, and S. LORENZ.) Über Abtötung von Krebszellen durch Röntgenstrahlen. *Z. Naturforsch.* **18b**, 654.

[455] 1963. (D. KAYSER.) Über den Mechanismus der Abtötung von Ascites-Krebszellen durch *Clostridium butyricum*. *Z. Naturforsch.* **18b**, 748.

[456] 1963. (With G. KRIPPAHL, K. JETSCHMANN, and A. LEHMANN.) Chemie der Photosynthese. *Z. Naturforsch.* **18b**, 837.

[457] 1963. (With P. OSTENDORF.) Quantenbedarf der Photosynthese in Blättern. *Z. Naturforsch.* **18b**, 933.

[458] 1963. Weiterentwicklung der zellphysiologischen Methoden. *Z. Klin. Chem.* **1**, 33.

[459] 1963. (With K. GAWEHN, A.-W. GEISSLER, and S. LORENZ.) Über Heilung von Mäuse-Ascites-Krebs durch D-und-L-Glycerinaldehyd. *Z. Klin. Chem.* **1**, 175.

[460] 1963. (With G. KRIPPAHL.) Über den Einfluss des Kohlensäuredrucks auf den Quantenbedarf der Photosynthese. *Acta chem. scand.* **17**, Suppl. 1, 1.

[461] 1963. (With G. KRIPPAHL.)Die chemischen Gleichungen der Photosynthese. *Hoppe-Seyler's Z. physiol. Chem.* **332**, 225.

[462] 1964. (With G. KRIPPAHL and E. BIRKICHT.) Quantenchemie der Photosynthese, automatisch geschrieben. *Biochem. Z.* **340**, 1.

[463] 1964. (With G. KRIPPAHL.) Uber das Verhalten von Asparaginsäure und Alanin in *Chlorella*, untersucht mit optischen Testen. *Biochem. Z.* **340**, 471.

[465] 1964. (With K. GAWEHN, A.-W. GEISSLER, D. KAYSER, and S. LORENZ.) The mechanism of biological X-rays action. *Naturwissenschaften* **51**, 373.

[466] 1964. Berichtigung. Bemerkung zu dem Nachruf von W. Kroebel auf James Franck. *Naturwissenschaften* **51**, 550.

[467] 1964. Prefatory-chapter. *A. Rev. Biochem.* **33**, 1.

[468] 1964. (With G. KRIPPAHL, K. JETSCHMANN and A. LEHMANN.) Chimie de la photosynthèse. *Bull. Soc. Chim. Biol.* **46**, 9.

[469] 1964. (D. KAYSER.) Über die Wirkung von Chinonen auf Ascites-Krebszellen *in vivo* and *in vitro*. *Z. Naturforsch.* **19b**, 258.

[470] 1964. (With E. BIRKICHT, G. KRIPPAHL, and A. LEHMANN.) New methods in measuring 1-quantum reaction of

photosynthesis. *Ber. Bunsen Ges. Phys. Chem.* **68,** 767.

[471] 1965. (With K. GAWEHN, A.-W. GEISSLER, D. KAYSER and S. LORENZ.) Experimente zur Anaerobiose der Krebszellen. *Klin. Wschr.* **43,** 289.

[472] 1965. (With G. KRIPPAHL and K. JETSCHMANN.) Widerlegung des Photolyse des Wassers und Beweis der Photolyse der Kohlensäure nach Versuchen mit lebender Chlorella und den Hill-Reagentien Nitrat und $K_3Fe(CN)$. *Z. Naturforsch.* **20b,** 993.

[473] 1965. (With A.-W. GEISSLER and S. LORENZ.) Messung der Sauerstoffdrucke beim Umschlag des embryonalen Stoffwechsels in Krebs-Stoffwechsel. *Z. Naturforsch.* **20b,** 1070.

[474] 1965. GERHARD DOMAGK, *Dt. med. Wschr.* **90,** 1484.

[475] 1965. On the mechanism of photosynthesis. *Agrochimica* **9,** 105.

[476] 1966. (With G. KRIPPAHL.) Über das Einsteinsche Gesetz und die Funktion des roten Ferments bei der Photosynthese. *Biochem. Z.* **344,** 103.

[477] 1966. *Über die letzte Ursache und die entfernten Ursachen des Krebses.* Verlag K. Triltsch, Würzburg

[478] 1966. Causes of cancer. *Colloq. Ges. Physiol. Chem.* **17,** 1.

[479] 1966. (D. KAYSER.) Über die Wirkung von Röntgenstrahlen auf das Wachstum von Lactobacillus Delbrückii und über die Rolle von Metallspuren beim Zustandekommen der biologischen Röntgenstrahlenwirkungen. *Z. Naturforsch.* **21b,** 167.

[480] 1966. (With A.-W. GEISSLER and S. LORENZ.) Irreversible Erzeugung von Krebsstoffwechsel in embryonalen Mäusezellen. *Z. Naturforsch.* **21b,** 707.

[481] 1966. Oxygen, the creator of differentiation. *In Current aspects of bioenergetics* (ed. N. O. Kaplan and E. P. Kennedy) p. 103. Academic Press, London.

[482] 1967. (With A.-W. GEISSLER and S. LORENZ.) Bemerkung über die Tryptophan-Oxygenase. *Hoppe-Seyler's Z. physiol. Chem.* **348,** 899.

[483] 1967. Neue manometrische Zweigefässmethode. *Hoppe-Seyler's Z. physiol. Chem.* **348,** 1677.

[484] 1967. (With A.-W. GEISSLER and S. LORENZ.) Wirkung von Riboflavin und von δ-Aminolävulinsäure auf wachsende Krebszellen *in vitro. Hoppe-Seyler's Z. physiol. Chem.* **348,** 1683.

[485] 1967. (With A.-W. GEISSLER and S. LORENZ.) Über Wach-

stum von Krebszellen in Medien, deren Glucose durch Galaktose ersetzt ist. *Hoppe-Seyler's Z. physiol. Chem.* **348**, 1686.

[486] 1967. (With E. BIRKICHT and R. STEVENS.) Quantenbedarf der Bilanz von Lichtreaktion und Rückreaktion bei der Photosynthese, automatisch geschrieben. *Biochem. Z.* **346**, 407.

[487] 1967. (With G. KRIPPAHL.) Photolyt und Chlorophyll in *Chlorella. Biochem. Z.* **346**, 418.

[488] 1967. (With G. KRIPPAHL.) Photolyt und Mangan in *Chlorella. Biochem. Z.* **346**, 429.

[489] 1967. (With G. KRIPPAHL and A. LEHMANN.) Über Photolyt und Phosphat in *Chlorella. Biochem. Z.* **346**, 434.

[490] 1967. *Gedenkworte für Gerhard Domagk.* Verlag Lambert Schneider, Heidelberg.

[491] 1968. (With G. KRIPPAHL and A. LEHMANN.) Metaphosphate and aerobic CO_2 in *Chlorella. FEBS Lett.* **1**, 171.

[492] 1968. (With A.-W. GEISSLER and S. LORENZ.) Wirkung von Riboflavin und Luminoflavin auf wachsende Krebszellen. *Z. Klin. Chem.* **6**, 467.

[493] 1968. (With G. KRIPPAHL and A. LEHMANN.) Dialysierte *Chlorella*, ein neues Versuchsmaterial zur Untersuchung der Photosynthese. *Z. Naturforsch.* **23b**, 1076.

[494] 1969. *The prime cause and prevention of cancer.* English edition by D. Burk. Verlag K. Triltsch, Würzburg.

[495] 1969. (With G. KRIPPAHL and A. LEHMANN.) Über die Chlorophyll-Katalyse der Photosynthese. *Z. Naturforsch.* **24b**, 1355.

[496] 1969. (With G. KRIPPAHL and A. LEHMANN.) Über das mit Kohlensäure verbundene und das freie Chlorophyll bei der Photosynthese. *Z. Naturforsch.* **24b**, 1583.

[497] 1969. (With G. KRIPPAHL and A. LEHMANN.) Chlorophyll catalysis and Einstein's law of photomechanical equivalence in photosynthesis. *Am. J. Bot.* **56**, 961.

[498] 1969. (With G. KRIPPAHL and A. LEHMANN.) Chlorophyll catalysis and Einstein's photochemical law in photosynthesis. *FEBS Lett.* **3**, 221.

[499] 1970. (With E. BIRKICHT and G. PAHLKE.) Änderung des Chlorophyllspektrums bei Bildung und Zerfall des Photolyten. *Z. Naturforsch.* **25b**, 112.

[500] 1970. (With A.-W. GEISSLER and S. LORENZ.) Entstehung

von Krebsstoffwechsel durch Vitamin-B_1-Mangel (Thiaminmangel). *Z. Naturforsch.* **25b**, 332.

[501] 1970. (With A.-W. GEISSLER and S. LORENZ.) Über die Erzeugung von normalem Stoffwechsel aus Krebsstoffwechsel durch Zusatz von Vitamin-Bs. *Z. Naturforsch.* **25b**, 559.

[502] 1970. (With E. BIRKICHT, G. KRIPPAHL and G. PAHLKE.) Changes of the chlorophyll spectrum in living *Chlorella*, if the photolyte is split by light. *FEBS Lett.* **8**, 247.

[503] 1970. Biologische Manometrie. In *Methoden der enzymatischen Analyse* (ed. H.-U. Bergmeyer), p. 208. Verlag Chemie, Weinheim.

Books

[504] 1926. *Über den Stoffwechsel der Tumoren.* Verlag Springer, Berlin. Translated into English by F. Dickens under the title *The metabolism of tumours.* Constable, London, 1930. (A collection of original papers published previously in various journals, prefaced by a chapter on methodology.)

[505] 1928. *Über die katalytischen Wirkungen der lebendigen Substanz.* Verlag Springer, Berlin. (A collection of papers published previously in various journals, prefaced by a chapter summarizing the work as a whole.)

[506] 1946. *Schwermetalle als Wirkungsgruppe von Fermenten.* Verlag Saenger, Berlin. Translated by A. Lawson into English under the title *Heavy metals and enzyme action.* Clarendon Press, Oxford (1949). (A review of Warburg's work on the subject and a discussion of related work by other authors.)

[507] 1949. *Wasserstoffübertragende Fermente.* Editio Cantor GMBH, Freiburg i. Br. (A collection of papers published previously, mainly in *Biochem. Z.*, prefaced by two extensive review articles).

[508] 1962. *Weiterentwicklung der zellphysiologischen Methoden, angewandt auf Krebs, Photosynthese und Wirkungsweise der Röntgenstrahlen.*

New methods of cell physiology, applied to cancer, photosynthesis, and mechanism of X-ray action. George Thieme Verlag, Stuttgart and Interscience, New

York. (This book contains an introduction summarizing the work in both German and English. In addition it contains a biographical sketch of Otto Warburg by D. Burk and 94 separate papers, most of which have been published elsewhere, with the following exceptions:

(a) (With G. KRIPPAHL.) Manometrische Bestimmung der Ascorbinsäure, p. 541.
(b) (With A. LEHMANN and H. W. GATTUNG.) Über den Quantenbedarf der Photosynthese in Blättern, p. 542.
(c) (With G. KRIPPAHL.) Weiterentwicklung der manometrischen Methoden, p. 555.
(d) (T. TERRANOVA and K. GAWEHN.) Stoffwechsel der Zellen des trypsinisierten Knochenmarks, p. 557.
(e) (With G. KRIPPAHL.) Ortho- und meta-Phenanthrolin in *Chlorella*, p. 560.
(f) (With K. GAWEHN and T. TERRANOVA.) Weitere Versuche über die Umwandlung des embryonalen Stoffwechsel, p. 562.
(g) (With A.-W. GEISSLER and S. LORENZ.) Neue Methode zur Bestimmung der Kohlensäuredrucke über Bicarbonat-Carbonatgemischen, p. 578.
(h) (With A.-W. GEISSLER and S. Lorenz.) Quantenausbeute der Photosynthese als Funktion des Kohlensäuredrucks, p. 582.
(i) (With A.-W. GEISSLER and S. LORENZ.) Weitere Versuche über die Wirkung von Röntgenstrahlen auf den Stoffwechsel der Krebszellen, p. 585.
(j) (With D. KAYSER.) Über die Wirkung der Röntgenstrahlen auf den Stoffwechsel von Milchsäurebakterien, p. 586.
(k) (With G. KRIPPAHL, H. GEWITZ, and W. VÖLKER.) Über Sauerstoffübertragung durch belichtetes Chlorophyll, p. 595.

INDEX

1 Emil Warburg (1846–1931), Otto's father.

2 Otto Warburg, *circa* 1951.

Jeder zeiget sich mir heute
Von der allerbesten Seite
Und von nah und fern die Lieben
Haben rührend mir geschrieben
Und mit allem mich beschenkt
Was sich so ein Schlemmer denkt –
Was für den bejahrten Mann
Noch in Frage kommen kann
Alles naht mit süssen Tönen
Um den Tag mir zu verschönen,
Selbst die Schnorrer ohne Zahl
Widmen mir ihr Madrigal.
Drum gehoben fühl' ich mich
Wie der stolze Adlerich.
Nun der Tag sich naht dem End'
Mach ich euch mein Kompliment
Alles habt ihr gut gemacht
Und die liebe Sonne lacht.

A. Einstein
peccavit 14 III 29.

Herzlichen Dank für die
freundlichen Wünsche und

die Blumen. Bei einem
anderen wüsst ich schon
fast nimmer, wie er ausschaut,
wenn ich ihn solange nicht
gesehen hätte.

Frohe Arbeit und herzlichen Gruss

Ihr
A. Einstein.

3 Letter from Einstein to Warburg acknowledging congratulations on his fiftieth birthday.

4 The Kaiser Wilhelm Institute for Cell Physiology, Berlin-Dahlem.

5 The Warburg apparatus.

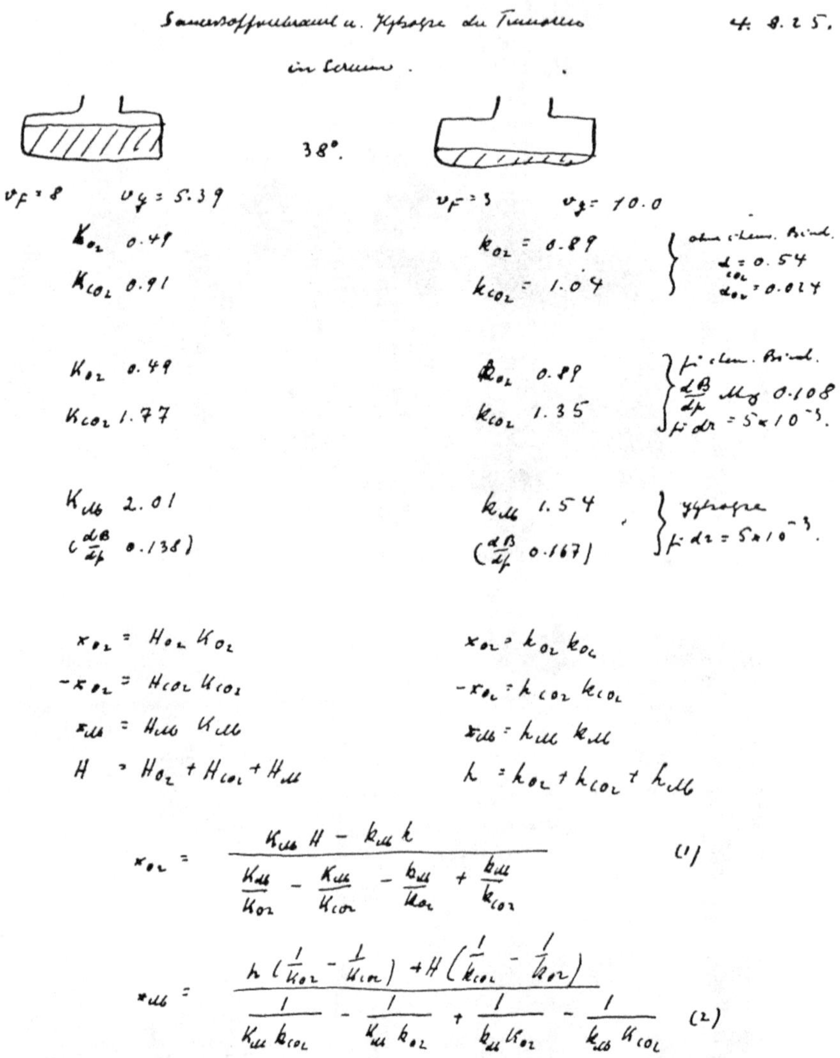

Sauerstoffverbrauch u. Glykolyse der Tumorzellen in Serum. 4. 8. 25.

38°.

$v_F = 8$ $v_G = 5.39$

K_{O_2} 0.49

K_{CO_2} 0.91

K_{O_2} 0.49

K_{CO_2} 1.77

K_{Mil} 2.01

$\left(\frac{dB}{dp}\right.$ 0.138$\left.\right)$

$v_F = 3$ $v_G = 10.0$

$k_{O_2} = 0.89$

$k_{CO_2} = 1.04$

ohne chem. Bind. $\alpha_{CO_2} = 0.54$, $\alpha_{O_2} = 0.024$

k_{O_2} 0.89

k_{CO_2} 1.35

mit chem. Bind. $\frac{dB}{dp}$ Mg 0.108, für $dr = 5 \times 10^{-3}$.

k_{Mil} 1.54

$\left(\frac{dB}{dp}\right.$ 0.167$\left.\right)$

Glykolyse für $dr = 5 \times 10^{-3}$.

$x_{O_2} = H_{O_2} K_{O_2}$

$-x_{O_2} = H_{CO_2} K_{CO_2}$

$x_{Milchs.} = H_{Milchs.} K_{Milchs.}$

$H = H_{O_2} + H_{CO_2} + H_{Mil}$

$x_{O_2} = h_{O_2} k_{O_2}$

$-x_{O_2} = h_{CO_2} k_{CO_2}$

$x_{Milchs.} = h_{Mil} k_{Mil}$

$h = h_{O_2} + h_{CO_2} + h_{Milchs.}$

$$x_{O_2} = \frac{K_{Milchs.} H - k_{Mil} h}{\frac{K_{Milchs.}}{K_{O_2}} - \frac{K_{Mil}}{K_{CO_2}} - \frac{k_{Mil}}{k_{O_2}} + \frac{k_{Mil}}{k_{CO_2}}} \quad (1)$$

$$x_{Milchs.} = \frac{h\left(\frac{1}{K_{O_2}} - \frac{1}{K_{CO_2}}\right) + H\left(\frac{1}{k_{CO_2}} - \frac{1}{k_{O_2}}\right)}{\frac{1}{K_{Mil} k_{CO_2}} - \frac{1}{K_{Mil} k_{O_2}} + \frac{1}{k_{Mil} K_{O_2}} - \frac{1}{k_{Milchs.} K_{CO_2}}} \quad (2)$$

6 Page from Warburg's laboratory notebook. This volume contains notes for the years 1925–45. Page 39 for 4 August 1925, reproduced here, relates to work on the manometric measurement of cell metabolism in serum published in *Biochem. Zeitschrift* **164**, 481, 1925.

7 Warburg's Nobel Prize document.

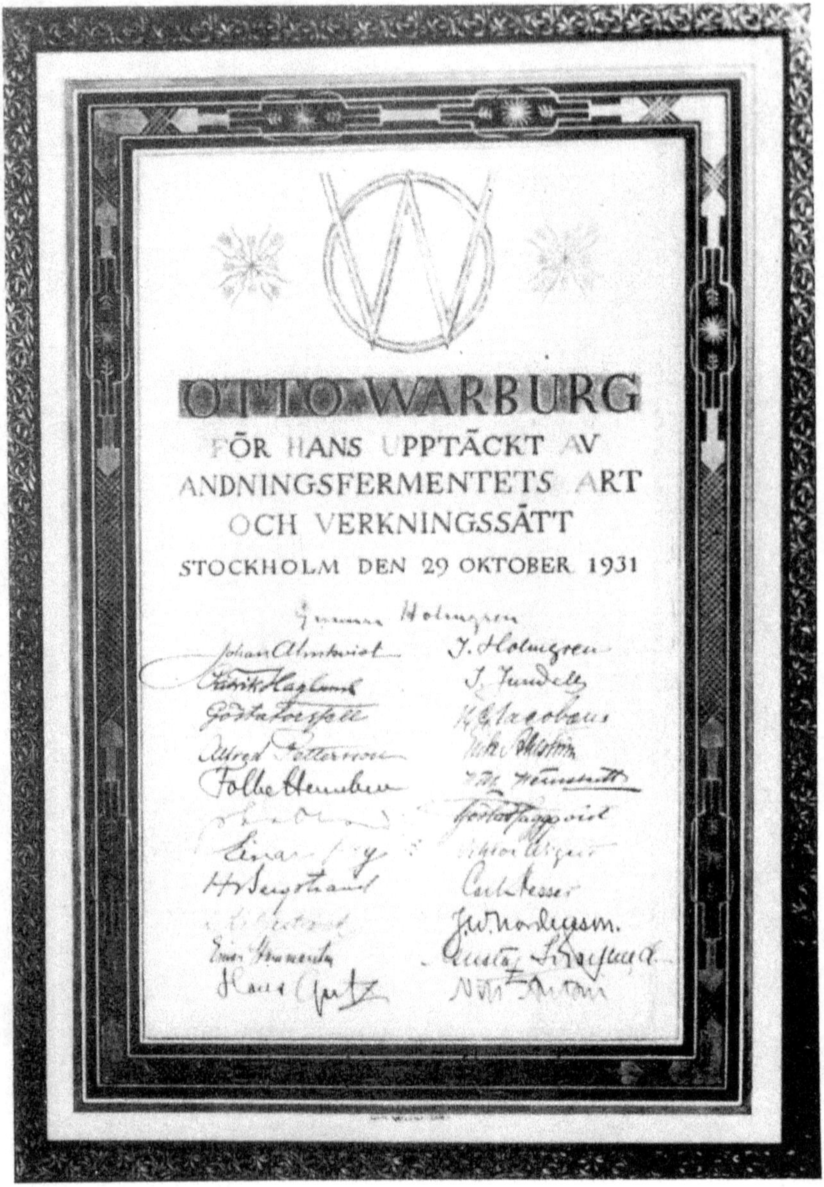
OTTO WARBURG
FÖR HANS UPPTÄCKT AV
ANDNINGSFERMENTETS ART
OCH VERKNINGSSÄTT
STOCKHOLM DEN 29 OKTOBER 1931

8 Prince Wilhelm and Otto Warburg, Stockholm 1931.

9 Otto Warburg in his laboratory.

10 Otto Warburg and Dean Burk.

11 Otto Warburg riding his dapple grey mare.

12 A winter stroll in the snow: Otto Warburg and Jakob Heiss with the poodle Bärchen and the boxer Carlo in 1960.

13 Meeting of Nobel Prizewinners, London, July 1963 (*see key below*).

KEY TO PLATE 13 (*above*)

1 T. H. Weller (Medicine 1954)
2 O. H. Warburg (Medicine 1931)
3 F. M. Burnett (Medicine 1960)
4 O. Hahn (Chemistry 1944)
5 P. S. Hench (Medicine 1950)
6 A. Butenandt (Chemistry 1939)
7 A. H. T. Theorell (Medicine 1935)
8 Count Lennart Bernadotte
9 S. Ochoa (Medicine 1959)
10 C. J. F. Heymans (Medicine 1938)
11 W. Forssmann (Medicine 1956)
12 M. Born (Physics 1954)
13 H. A. Krebs (Medicine 1953)

14 Otto Warburg and Hans Krebs at the meeting of Nobel Laureates, Lindau, June 1966.

15 At a Guinness Symposium in Dublin, September 1965: A. K. Mills, Hans Krebs, Otto Warburg, and Feodor Lynen.

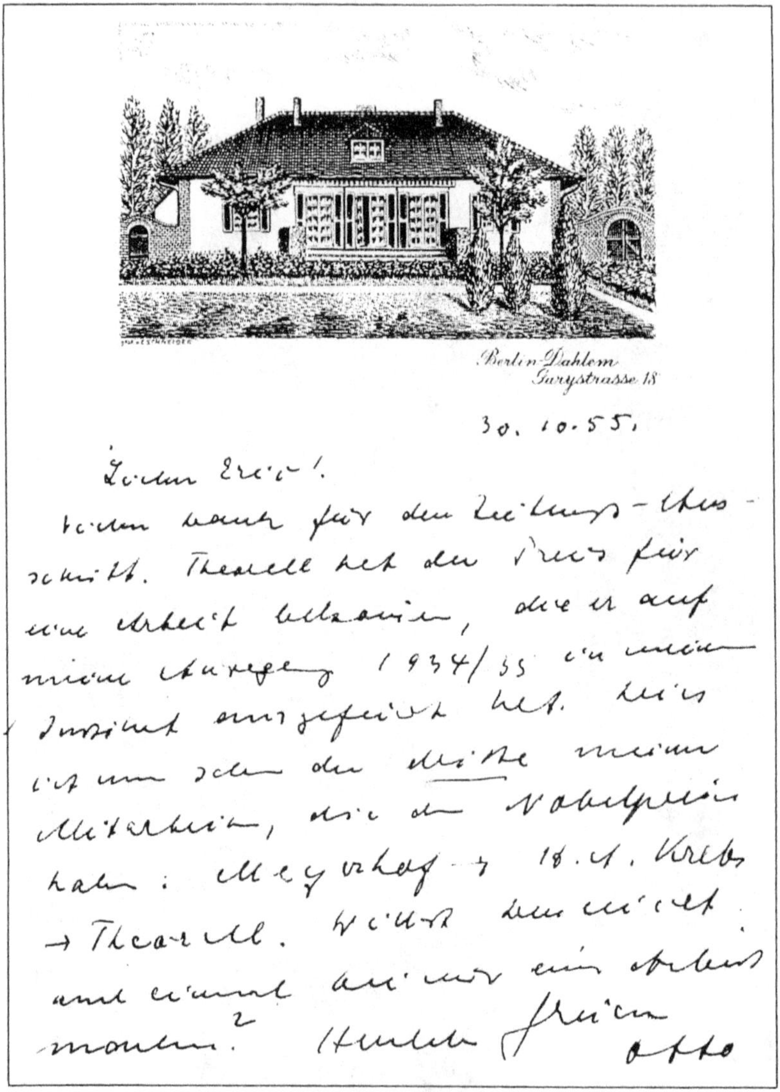

Berlin-Dahlem
Garystrasse 18

30. 10. 55.

Lieber Eric!

Vielen Dank für den Zeitungs-Ausschnitt. Theorell hat den Preis für eine Arbeit bekommen, die er auf meine Anregung 1934/35 in meinem Institut ausgeführt hat. Dies ist nun schon der dritte meiner Mitarbeiter, die den Nobelpreis haben: Meyerhof, H. A. Krebs, Theorell. Willst Du nicht auch einmal wieder eine Arbeit machen?

Herzlichst Dein Otto

16 Postcard from Otto Warburg to Eric Warburg, 10 October 1955. The engraving shows Warburg's house at Dahlem.

17 Warburg at 18 Garystrasse with his bust by the sculptor Richard Scheibe.

18 Warburg on his seventieth birthday (8 October 1953) holding the Grosse Verdienstkreuz (Grand Service Cross) with star.

19 Professor Richard Kuhn presents Otto Warburg, on his seventy-fifth birthday, with the certificate of the honorary Doctorate of the Faculty of Medicine, University of Heidelberg (8 October 1958).

20 Portrait in oils by Y. Oberland (1958) with order and decoration.

21 The Otto Warburg Medal. For the eightieth birthday (8 October 1963) of its Honorary Member the Gesellschaft für Biologische Chemie founded the Otto Warburg Medal (bronze by Richard Scheibe). The prize, which consists of a medal and a sum of money, has been awarded annually since 1963: recipients include Feodor Lynen (1963); Kurt Mothes (1965); Michael Sela (1968); Hans A. Krebs and Carl Martius (1969); Ernst Klenk (1971); Hans Leo Kornberg (1973); Theodor Bücher (1974); Helmut Holzer (1975); Heinz-Günter Wittmann (1976); Robert Huber (1977); Wilhelm Stoffel (1978).

22 Otto Warburg in the laboratory, July 1966.

CPSIA information can be obtained
at www.ICGtesting.com
Printed in the USA
BVHW052239161019
561270BV00006B/201/P

9 784871 871525